Encyclopedia of Professionalization

SCIENCES

Education and Training, Field Director – Jean-Marc Labat

Education and Society,
Subject Heads – Géraldine Farges and Xavier Pons

Encyclopedia of Professionalization

Organization of Professions, Production of Professionalities and Growth of Professionalism

Coordinated by
Didier Demazière
Richard Wittorski

ISTE

WILEY

First published 2024 in Great Britain and the United States by ISTE Ltd and John Wiley & Sons, Inc.

ISTE Ltd
27-37 St George's Road
London SW19 4EU
UK

www.iste.co.uk

John Wiley & Sons, Inc.
111 River Street
Hoboken, NJ 07030
USA

www.wiley.com

Library of Congress Control Number: 2024943305

British Library Cataloguing-in-Publication Data
A CIP record for this book is available from the British Library
ISBN 978-1-78945-183-2

ERC code:
SH3 The Social World and Its Diversity
SH3_8 Social policies, welfare, work and employment
SH3_11 Social aspects of teaching and learning, curriculum studies, education and educational policies

Contents

Introduction

Didier DEMAZIÈRE[1] and Richard WITTORSKI[2]
[1] *CSO, CNRS, Sciences Po, Paris, France*
[2] *CIRNEF Laboratory, University of Rouen, France*

Professionalization is a concept with a prominent presence in the fields of education, training and work, as well as in public and private organizations. Many observers agree that the term has become polysemous, as its uses dramatically vary depending on the actors using it and their various reasons: to have professional experience recognized, to adapt training to the productive system, to team up in peer communities to improve their status, to invite people to engage in new forms of work, etc. In addition, polysemy around this topic has given rise to debates and tensions as to the different interests invested in the term's use.

Briefly stated, within organizations, professionalization can denote the appearance of new work expectations: in training spaces, it can translate learning processes and the construction of knowledge and skills (a term often used when we talk about professionalization); in public policies, it can express intentions for reforms and the establishment of new forms of governance; it can also be used by employees and professional groups or unions to foster the recognition of activities, professions, statuses, etc.

The notion of professionalization may alternately respond to a quest on the part of employees and "tradespeople" for further recognition or to the rise in demands addressed to employees or professionals by organizations, clienteles or users around efficiency challenges or work quality. These two movements do not necessarily converge, and rarely do they reflect the same expectations or practices. Lexical

Encyclopedia of Professionalization,
coordinated by Didier DEMAZIÈRE and Richard WITTORSKI.

consensus quickly gives way to strong semantic dissent leading to discrepancies, misunderstandings and even tensions.

Now we observe that professionalization can refer to various phenomena. As a result, we may notice a plethora of analytical grids across research papers whose unity is difficult, if not impossible, to grasp. Therefore, professionalization is located at the crossroads of multiple disciplinary perspectives: the educational sciences, labor sciences, organizational sciences and political sciences. From a historic viewpoint, the concept of professionalization was first theoretically developed by sociology in English-speaking countries. This discipline asserted itself with the booming of the "sociology of professions" specialization, which studies the emergence and constitution of professional groups defending their position in the division of labor. In broad terms, the sociology of professions in English-speaking countries gradually evolved after the beginning of the 20th century. Sometimes taking a stand in the field of social and political debates, it intended to report on the way in which groups of individuals sharing the same activities organized themselves in free market contexts in view of obtaining an authorization to exercise, conquer a monopoly of exercise and regulate the professional territory conquered. A heated debate quickly crossed this field opposing functionalists to interactionists, while the former sought to elucidate the distinctive traits enabling and legitimizing the access to the status of *profession* (as opposed to the common condition of *occupation*), the latter rejected said distinction and attempted to explain why certain activities manage to be recognized as professions, arguing for the contingent nature of the difference. Other Weberian, neo-Weberian and Marxist approaches later rekindled this debate.

In countries with strong state presence and central regulation of activities (mainly the old continent countries), the collective challenges are not the same and the recognition dynamics of professional activities do not engage the same logics, nor do they follow the same avenues. Work is being carried out in Europe, on the one hand, in reaction to the approaches prevailing in English-speaking countries – deemed poorly valid for characterizing the vast majority of professional activities which do not fall within liberal professions – and, on the other hand, in continuity with an interactionist perspective by focusing on the internal and external dynamics of professional groups, or by developing a "conflictualist" approach to relationships between professional groups and institutions or organizations. In the wider field of sociology of work, notably in France, Belgium, etc., numerous "critical" papers have studied the development and use of the professionalization glossary by companies (the terms *skill*, *performance*, *autonomy*, *responsibility*, etc.) in connection with organization, regulation and work assessment challenges. Some have insisted on the idea that professionalization joins a growing call for skills to meet new work standards. The challenges could involve "making people swallow the pill of flexibility". Permanent adaptability could lead to the individualization of assessment, tightening control over workers and reducing their autonomy at work. From this point of view, professionalization is also perceived as being at the service of committing employees to more flexible work contexts and making greater use of people's subjective resources.

Recently, the field of training has also adopted the term, due to its massive introduction since the end of the 1990s into international, European and national discourses and texts governing professional training, whether initial or ongoing. Essentially, the work carried out there concerns the analysis of trends in the joint evolution of work and training, and the identification of new work-training articulation logics in so-called "professionalizing training" approaches, more closely related to changes in work and supposed to prepare the trainee for more targeted functions.

As we can see, the concept of professionalization has been subject to fragmented theoretical development, in different scientific fields at the origin of varied research traditions. However, work challenges, wider political and social challenges, challenges raised by professional groups and training challenges are closely intertwined when discussing professionalization.

To speak of professionalization in the field of training necessarily implies evoking the transformations of work and broader environments at the same time, seeking to understand the reciprocal transformations of individuals, collectives, organizations and environments. The general hypothesis could be the following: training and learning processes, as well as their transformations, can only be understood located in "space-time" (socio-political, organizational and collective environments (professional groups and local micro-collectives), individuals), themselves characterized by reciprocal transformations encouraged by inseparable challenges in the continuous change of organizations, collectives and individuals. By extension, in order to understand these dynamics, it is necessary to cross-research traditions that are rarely linked: for example, those centered on the transformations of legal regulations, the dynamics of organizations, the collective of workers, activity analyses, the metamorphosis of knowledge, the construction of the individuals' experience and career path, the configuration of individual and collective identities, etc.

The ambition of this encyclopedic volume is to bring together and stimulate the dialogue between the available research papers addressing this concept in various fields. In fact, this insight cannot forgo the different "theoretical lenses" and the knowledge they produce because, as we mentioned earlier, the dynamics of professionalization are inseparable from ampler organizational, sociopolitical and collective dynamics. Furthermore, the knowledge produced by these approaches is today difficult to gather because it is dispersed across various scientific fields and traditions: the sociology of professions, work analysis (sociology of work and psycho-sociology of work), adult education and training, management sciences and political sciences, to mention the most important ones. As a corollary, this dispersion leads to low visibility and a difficulty for illuminating the field of professionalization. This ambition is reinforced by the idea that it is necessary to shed light on the challenges of professionalization for education and training by relating them to political and social challenges, organizational challenges, collective and individual challenges.

To satisfy these intentions, this encyclopedic volume will be organized into three parts which correspond to three main orientations or social significations of professionalization, supplemented in each case by texts penned by one or more specialists in the area concerned.

The first part will address the issue of professionalization probably in its oldest sense: the emergence, the constitution, the quest for recognition and the hoped for or effective institutionalization of professions. The challenge is to empirically grasp and theoretically model the professionalization paths of professional groups, of workers carrying out a given activity, seeking to have it recognized and valued, to gain autonomy in the regulation of their work with regard to their environment (clientele, organizations), or even to establish a monopoly of exercise ensuring them protection against competition from other professionals: which groups succeed in achieving this and how? What are the paths to professionalization? How can we account for and explain the considerable inequalities in this area? Showing that the sociologists of professions and those of work have provided divergent answers to these questions, two texts explore the concept of professionalization as "the manufacturing of professions". Florent Champy will explore the potential of a "prudentialist" theory of professionalization, showing that it informs the historical process of protection of certain activities and sheds light on contemporary managerial threats risking to degrade the services provided by these professions. Didier Demazière will favor the prism of the boundaries dynamics delimiting professional groups, arguing that such boundaries are invariably shifting and unstable because professionalization is always uncertain, so that it can designate the division of labor and the recomposition of activities, which comes down to stating that professional groups are professionalizations.

In the second part, professionalization will be studied as the "ascent of professionalism", that is to say, as the new requirements at work, the recent standards to define and supervise it, and which contribute to recognizing the employee as a professional (or not). New ways of defining, regulating and evaluating work are emerging, often associated with a new rhetoric of quality and efficiency, which ultimately raises the question as to who "is in control": the professionals or the organizations, the customers or the institutions which finance the activity and which then impose them on the professionals? Or are they the subject of negotiation, and if so, between which actors and with what outcomes? This part will be devoted to the second meaning of the term professionalization, which will be discussed in three texts: firstly, Arnaud Mias will address the reasons and forms underlying the emergence of new requirements at work. Valérie Boussard will then discuss the way in which these new requirements bring to the forefront new conceptions of a job well done, sometimes in tension with those held by professionals. Mokhtar Kaddouri will finally show that professionalization is part of a debate between organizational, collective and individual projects, more lively than ever when new requirements are imposed on professionals.

In the third part, professionalization will be discussed from the perspective of the "production of professionalities", in other words, the preparation of future professionals or the further development of "existing" professionals not only through training but also through work, activity and the construction of experience. The goal is to study professionalization from the angle of the acquisition and development of expertise and knowledge specific to particular areas of intervention. In this sense, professionalization often (but not only) involves education and training systems, as well as broader work and social life experience. At this point, we can see that social expectations enlarged after the international and European provisions of the late 1990s, which required education and training systems to further professionalize their supply. The texts by Philippe Maubant, Stephen Billett, Sandra Enlart, Lindon West and Joris Thiévenaz will show that the production of professionalities goes far beyond the recourse to the "typical" apprenticeship program (*formation par alternance*), often giving a preponderant place to the field of training in the organizational decisions of such schemes. The approach will diversify and become more complex with the need to acknowledge the learning nature of the professional activity itself (like on-the-job training actions developed in France), as well as the work contexts/environments, which all lead to the production of professionalities also as a form of professional socialization.

In a final text, Richard Wittorski will discuss the hypothesis that the three meanings of professionalization which structure the encyclopedia, far from being separated and functioning autonomously, are in reality strongly articulated, revealing particular configurations, that is to say, close interdependence systems. He will develop the idea that professionalization points to “space-time” dimensions or singular configurations (forms) articulating (putting into debate-tension, with different combinations) the projects of professional groups, organizations and individuals in (socio-politico-economic) environments which have their own characteristics.

PART 1

Professionalization, the Manufacturing of Professions

1

Professionalization, Practical Wisdom and Vulnerabilities

Florent CHAMPY
LISST, CNRS, University of Toulouse, France

1.1. Introduction

Professionalization – to be understood as achieving a status protecting certain activities from competition in markets and from power in organizations – is a central theme in the sociology of professions[1]. It is also a process which mainly took place from the middle of the 19th century to the middle of the 20th century. The history of medicine is emblematic of this process. How can we account for the statuses granted to certain professions? Sociologists have offered contradictory answers to this question, and even the most outstanding among them (those provided by functionalists, interactionists and Andrew Abbott) have led to aporias.

The Aristotelian concept of *phronesis* enables us to overcome these difficulties. Translated into English as "practical wisdom", it denotes a mode of thinking required for acting in situations of irreducible uncertainty (Aubenque 1963; Broadie 1991). Practical wisdom both explains the fragility of protected professions (which

1 This chapter was made possible thanks to discussions with various colleagues, although it is impossible to list them all. However, I would like to mention Julien Barrier, Teresa Carvalho, Patrick Castel, Tiago Correia, Lilian Gravière, Morgane Kuehni, Elodie Lemaire, Christian Mouhanna, Frédéric Pierru, Bertrand Ravon and Richard Wittorski. The original manuscript also greatly benefited from the insightful rereading by Olivier Pilmis. May they all be very warmly thanked.

are prudential[2] activities), and the challenges of not letting untrained persons replace professionals at work. It accounts for the fact that medicine (which philosophers considered to be the prudential activity *par excellence*) is the model for professionalization and the activity that has occupied sociologists to the largest extent. Briefly stated, a prudentialist theory of professionalization sheds light on the forms and consequences of recent managerial offensives to reduce the autonomy or advantages of such professions. It helps us to understand the fragility of activities especially targeted by a mode of thinking, dominant in society, which defies the requirements of practical wisdom.

To show this, we will examine the explanations that the three aforementioned trends have provided for professionalization and their limitations. We will then explore the arguments in favor of a prudentialist theory of professionalization, principally relying on an analysis of the dominant conceptions of professionalism at the end of the 19th century and early 20th century, when most professionalization took place. Finally, we will discuss a more recent past, marked by the hold of managerial logic on professional working environments to show that the pressures the prudential professions[3] are confronted with run contrary to the value of practical wisdom in the times of professionalization.

1.2. Classic professionalization theories

1.2.1. *The functionalist theory*

The theme of professionalization is consubstantial with the emergence of the functionalist sociology of professions. Indeed, the functionalists distinguished between professions and occupations, explaining their differences within the

2 Prudential is the adjective associated with the noun expression "practical wisdom", while "prudence" and "prudencia" are the usual translations from the Greek *phronesis* into Latin languages.

3 According to Aristotle and his commentators, any activity is capable of giving rise to practical wisdom. We might therefore be surprised that the latter serves to identify a particular type of activity. However, sociological work has uncovered several consequences of high-density practical wisdom, which are the distinctive features of said activities. Three will appear here: the need to strike a balance between the activity's conflicting goals, something which often leads to mobilizations to impose a specific conception of such equilibrium (Champy 2018); the theory of professionalization; the particular vulnerabilities addressed in Part 3. Actually, an accurate list of professions involving prudential practices is impossible to draw up. In certain activities, only one or two of these features are found. However, "*we cannot emphasize this enough:* it is not because their boundaries are unclear that the categories do not exist." (Héran 2021, p. 83).

framework of a theory of professionalization, showing what distinctive features had permitted certain activities to access the status of a profession, the advantages they derived from it and what social functions were performed by the protection that this professionalization process offered them. The functionalists worked on this theme from the beginning of the century until the end of the 1960s. As early as 1915, in a conference where he spoke about social work, Abraham Flexner presented the recourse to scientific knowledge and the tendency to self-organization as two features of professions. With slight variations, these two themes remained present throughout the half-century when functionalist theories dominated (Champy 2012).

The great functionalist theorist of professions was Talcott Parsons, who conceived society as a living organism within which professions constitute an organ (Parsons 1939, pp. 56–58). His sociology aimed to show that this organ is adapted to its function: by seeking to rationally cope with a series of social problems, professions contribute to the proper functioning of society, whereby professionalization becomes an element within a rationalization process. Parsons reflected upon professions as a whole, inspired by the case of medicine. By taking care of disease, medicine contributes to the functioning of the economic system. It is for this reason that the quality of the patient–doctor relationship has to be protected. Doctors benefit from significant protection in the form of a monopoly of practice, the control of teaching by professionals and strong autonomy in relation to non-doctors (only a doctor can legitimately judge the validity of a medical decision). While monopoly protects against competition in the labor market, autonomy prevents interference in organizations: professional logic is deployed in articulation/competition with the market and the organizational logics of economic life.

According to Parsons, two elements explained these forms of protection. To begin with, only doctors can treat the sick because they have mastered the high-level scientific knowledge indispensable for care: a certain autonomy must be granted to them. The emergence of professions is inseparable from the development of universities as a place of research and training in science for future professionals, as these professions constitute the vector for the application of science to concrete problems[4]. This means that scientific knowledge supports the instrumental rationality of professions. Furthermore, the doctors' work is defined and supervised by the profession, in such a way that doctors use the autonomy conferred on them in the patient's interest. This normative framework is inscribed in the "role" of the doctor, and the "Order" as an institution ensures its respect. In other words, the profession's collective autonomy is more relevant than that of each doctor.

---

4 The same idea is found in Abbott, but with completely different implications (see section 1.2.3).

Primarily inspired in the case of medicine, the Parsonian theory was then softened to account for the professionalization process in law and engineering (Parsons 1968, pp. 536–546), with variations depending on activities and countries (the United States, United Kingdom and continental European countries). Nuancing his 1939 reflections, Parsons then placed less emphasis on the strength of the link between the development of universities and professionalization: the case of engineering showed that professionalization could also be achieved based on practical knowledge, as an extension of craftsmanship (a consideration which weakened his initial theory). Furthermore, even for medicine, the explanation of protection via scientific knowledge was not thoroughly convincing. As we will come back to later, the scientific bases of medicine were still almost non-existent in the second half of the 19th century, when its professionalization process began. Robert Merton, another functionalist, emphasized the lack of seriousness of medical education in the 19th century (Merton 1957).

At length, in 1964, Wilensky published an article where he remarked that more and more occupations were trying to attain a professional status, thus translating the success of the ideas of professionalism/professionalization in the world of work. Wilensky was the last notable functionalist working on professions as he stuck to the differentiation between occupations and professions: while many activities aspired to professionalization, few managed to reach the type of protection that the functionalists sought to explain. According to Wilensky, the service activities which emerged at that moment were not to be confused with professions, because they did not share their distinctive features. His article therefore showed how a functionalist sociologist sought to take into consideration the social developments of their time. However, above all, it represented an attempt to respond to the interactionists, who had already begun to draw radical conclusions from general aspirations for the status of a profession, by denouncing the distinction between professions and occupations, at the cost of new difficulties for the theory of professionalization.

1.2.2. *The interactionist critique*

From the 1950s, sociologists gathered in Chicago, notably around the figure of Everett Hughes, and founded a new school of thought. Among this group, Anselm Strauss, Howard Becker and Eliot Freidson, along with Hughes himself, made an important contribution to the sociology of work and professions based on a virulent critique of functionalism which was dominant at the time. At the end of the 1960s, their approach prevailed. While medicine continued to play a large part in their work, the group's main innovation was to open up the sociology of professions to the study of a wide variety of occupations, which they considered to be just as interesting as professionalized professions:

> Specifically we need to rid ourselves of any concepts which keep us from seeing that the essential problems of men at work are the same whether they do their work in the laboratories of some famous institution or in the messiest vat room of a pickle factory. [...] perhaps there is as much to be learned about the high-prestige occupations by applying to them the concepts which naturally come to mind for study of people in the most lowly kinds of work as there is to be learned by applying to other occupations the conceptions developed in connection with the highly valued professions. [...] It is not that it puts one in the position to debunk [them], but simply that processes which are hidden in other occupations come more readily to view in these lowly ones (Hughes 1984, pp. 342–343).

We thus see that the disinterestedness of doctors is more of an ideology than a reality, whereas the difficulty of the problems solved by plumbers makes us aware of their high competence level. In a nutshell, for Hughes, the lack of distinction between professions and occupations was a methodological choice. Any ontological affirmation was still absent, though. On the contrary, Becker asserted that there is no difference between professions and occupations. According to him, the concept of profession is merely a "folk concept" (Becker 1962).

This methodological (and then conceptual) innovation had a major consequence on the study of professionalization. If there is no difference between *professions* and *occupations*, it is futile to try to explain why certain activities have attained the status of profession, while others have not. Interactionists either avoided the question altogether (Hughes's position), or they presented professional status as the contingent result of struggles due to the advantages it provides (ibid.). In both cases, they dismissed any search for the bases of the distinction between professions and other occupations. However, by asserting that professions are only activities that have had better fortune than others in a competition, Becker actually developed a theory of professionalization which must be examined as such. In fact, this position – which has generated an extensive legacy, especially in the French sociology of professional groups – is not tenable.

By neglecting to explain why certain activities have managed to benefit from a professionalization process, while others have not, and simply by relying on ad hoc historical explanations, Becker avoided the pitfall of the functionalist theory: its difficulty in stabilizing the features that explain why an activity has reached the status of a profession, and in formalizing a typically relevant process in the face of the diversity of concrete historical processes. However, the cost was high. Not only did invoking chance or luck amount to purely and simply renouncing any explanation of the status granted to certain activities, but this assumption on chance had been invalidated by the facts, as it failed to account for a strong regularity: the same activities were protected in various countries. Despite variations, medicine is protected in all countries with a university system and a regulatory state, while other activities, like artisanal or commercial activities, are not protected anywhere. This regularity would be compatible with a chance-based explanation only if the professionalization process had taken place within a supranational framework, leading to the same results in different countries. However, this is not the case. Historians of professions agree on the importance of national frameworks to account for the progress and results of these processes (Malatesta 2006). In fact, these dissociated processes produce a result that is remarkably similar from one country to another. In any case, a theory of professionalization cannot avoid explaining this regularity.

The criticism of the implicit Beckerian theory of professionalization can also be formulated from within interactionism itself. In point of fact, Becker's ontological posture is not representative of this entire trend. Similarly to the functionalists before them, Strauss and Freidson placed medicine at the heart of their research and found unique features in it that they insisted upon, starting from the 1980s. Strauss then wrote:

> There are two striking features of health work shared only with certain other kinds of work. One consists of the unexpected and often difficult to control *contingencies* stemming not only from the illness itself, but also from a host of work and organizational sources as well as from biographical and life-style sources pertaining to patients, kin, and staff members themselves. A second and crucial feature of health work is that it is 'people work'. [...] taken together, both features insure that trajectory work harbors the potential for being complex and often highly problematic (Strauss et al. 1985, p. 9).

As for Freidson (1970, 1986), he was critical of professions, studying the misuses of professional power first in medicine, and then more generally. He showed that the scientific knowledge that doctors and members of the other professions invoke to justify their status and autonomy poorly reflects the reality of their activity, where many decisions lack a solid scientific basis, and may therefore appear arbitrary. This criticism is pertinent as it addresses a key theme both for professional rhetoric and functionalist theory: the ability of professionals to base what they do on fully objective reasons. However, Freidson never completely denied the scientific grounds of certain activities: for him, it was the gaps, that is to say, the effective application of knowledge, which posed a problem. Moreover, his later works pleaded for the defense of the professional model against the two other logics, market logic and organization logic (Freidson 1994, 2001), revealing his discomfort with the assertion of indistinctness between professions and occupations. His last works did not provide an original definition of professionalism that would enable us to avoid the pitfalls of functionalism. The quote by Strauss hints at a solution, suggesting that work is less objective than professionals claim and needs to be protected precisely because the complexity of situations invalidates the application of scientific knowledge. Andrew Abbott's theory is a more elaborate outline of this solution.

1.2.3. *Andrew Abbott's theory*

The last major theoretical work on professions is that by Abbott (1988), who provided an extremely stimulating theory of professionalization, acknowledging both the contributions of interactionism and those of functionalism. From the interactionists, Abbott took up the criticism of the "distinctive features" approach to professionalization, which he rightly criticized for having oversimplified reality. In contrast to the approach trying to update a professionalization model that was valid everywhere, Abbott claimed to be attentive to the diversity of concrete processes, recognizing within them a large part of contingency. Furthermore, he placed the concrete struggles between the actors and the division of labor at the heart of his theory: what professions are depends less on the adaptation of an organ to a function (as stated by the functionalists), than on their construction through multiple interactions inevitably involving an element of contingency. Finally, as the interactionists did, he studied more numerous and more diverse activities than those the functionalists were interested in. For these reasons, and also no doubt because by imposing its own research standards, interactionism established itself as a prism for reading other works, Abbott is often read as an interactionist, but this reading is incomplete, and the interest of his work also lies in the way in which he avoids the aporias of the interactionist perspective, notably the Beckerian perspective. Three points clearly differentiate him from the interactionists and bring him closer to the functionalists. First of all, Abbott again took up a question close to the functionalists' questions on professionalization, as he sought to discover how professions manage to obtain a lasting protection of their jurisdiction: behind the games of the actors, stability was of great interest to this author. Moreover, he broke with the perspective inherited from Hughes, which was to study every activity in the same way. Instead, he became interested in particular types of professions, those "applying somewhat abstract knowledge to particular cases"[5] (p. 8). Last, but not least, moving away from the idea of contingency that was dear to interactionists, he thoroughly explained the delimitation of jurisdictions and their protection through the concept of "efficacity", which held a crucial place in his theory. This book aims to take up the interactionist heritage (with the actors, struggles, strategies, contingency and the collective work for constructing the meaning of the social

5 Abbott's developments on this definition bring him close to functionalists, since it is the mastery of a body of abstract knowledge that is at stake, as with the doctors and engineers studied by Parsons.

world), while avoiding the radical relativist position which resulted from it and led to an aporia in the theory of professionalization.

How does Abbott manage to reconcile propositions which, in light of previous theories, seem contradictory? How can he speak both of the contingency of processes and the conformity of their results to efficiency requirements? The answer lies in the importance of temporalities in the struggle of profession members for certain jurisdictions, that is, to impose a relationship between a set of tasks and their profession. Abbott discerns between three arenas in which the efficiency of the response provided by a profession to a problem is recognized according to a very different temporality: the workplace, public opinion and the legal system (ibid., pp. 59–60). It is at the workplace that the division of labor adapts itself most quickly to efficiency requirements, because daily relationships prevent the lasting concealment of efficiency gaps between various solutions. When the members of an occupation in competition with others provide a more efficient solution to a problem, it will only take two or three years for this solution to be adopted. The image that professions give of their work to the lay public far from the workplace evolves at a slower pace: Abbott estimates the time it takes for this public to become aware of the change in the division of labor to be 10–20 years. Ultimately, the arena of the legislative or administrative system is even slower than that: it is only after 20–50 years that the administrative system formalizes any modification of the division of labor. The sluggishness of this process leaves room for power games, enabling some occupations to gain momentary advantages that the efficiency of their work does not justify: there is the place for the contingencies interactionists have insisted so much upon. However, according to Abbott, in the end efficiency is indeed the main factor which determines the sharing of jurisdictions, and which limits the contingent effects of the power games interactionists have emphasized[6].

Neither functionalist nor interactionist, this sophisticated theory describes how the division of labor and its recognition enshrined in law for certain professions end up conforming to what allows for the greatest efficiency of work; it contains a blind spot that its sophistication has long hidden. By describing *how* professionals gradually obtain the recognition of a jurisdiction in three successive arenas, it does not explain *why* the public authorities grant formal protection of this jurisdiction to some of them. Professionalization therefore remains unexplained. Furthermore, this blind spot leads to a contradiction: if the struggles at the workplace cause a new, more effective division of labor within a short period of time, as soon as an

6 For a more detailed presentation of this theory, see Champy (2012).

innovation justifies replacing the old one, any obstacle to competition between the members of different occupations can hinder this adaptution, and thus the efficiency. By protecting a profession's jurisdiction, public authorities freeze a division of labor that may not always be the most efficient. Not only does Abbott's theory not explain this formal protection, but the existence of such protection seems to invalidate the theory, as it can hinder efficiency, which plays a key role in this theory.

One solution, which is simple but not obvious, is that what is fragile and threatened and, for one reason or another, precious is secured. In fact, Abbott's book allows us to understand that professional jurisdictions are made vulnerable to competition by the difficulties that professional work faces, because it aims in "applying somewhat abstract knowledge to particular cases". For the activities Abbott was interested in, the application of abstract knowledge presents the risk of not sufficiently taking into consideration the complexity and singularity of the concrete cases discussed, thus facilitating the criticism of the work and the competition's claims over that market share. Besides, the very difficulties for applying knowledge to particular cases give reason to ensure that only qualified professionals can achieve this.

One of the originalities of this book is to propose an imposing theory of professional work, closely related to the theory of professionalization. According to Abbott, work has three dimensions: diagnosis, treatment and, between the two, inference, to which he devotes long passages. "Inference is undertaken when the connection between diagnosis and treatment is obscure" (ibid., p. 49). The inference chain serves to reflexively adapt the treatment to each unique case encountered. Indeed, this chain's length holds a significant place in the theory. If this chain is too short, the treatment can be inferred from the diagnosis and the profession's jurisdiction is fragile because this makes it easy for non-specialists to replace professionals: there is then a risk of an excessive mechanization of professional work. Conversely, if the chain is too long, the link between treatment and diagnosis becomes tenuous; therefore, the treatment seems purely subjective and no longer makes it possible to stabilize a jurisdiction. The activity then becomes vulnerable to interference from people outside the profession (e.g. the organization's management) and to jurisdictional claims by competitors. In other words, inference is used to search for an answer that cannot be obtained from objective deduction, but which cannot be contingent, either. Also, the jurisdiction can only be protected, provided that the work presents these qualities: it is not mechanical, but does not appear as totally contingent. Now these are precisely the qualities of practical wisdom, in the philosophical sense. Let us tackle that question.

1.3. The protection of professions with prudential practice

1.3.1. *The fragilities of practical wisdom and the theory of professionalization*

While the fragilities of practical wisdom seem to be able to justify the jurisdiction of prudential professions being formally protected, this is not the case for activities where practical wisdom plays a lesser role. Absent from Abbott's theory, this philosophical concept allows us to reflect upon this dead angle and escape from the aporias inherited from functionalism and interactionism. Before delving into what supports this thesis, let us explore what prudential professions are.

Practical wisdom is a mode of thinking that guides action in situations whose singularity and complexity entail irreducible uncertainty (Aubenque 1963). It is also a virtue, in that it aims to avoid the harm which could befall others if such uncertainty was ignored. Ultimately, this concept designates the dispositions and the aptitudes required to understand situations in which the mechanical application of abstract rules, formalized procedures, scientific knowledge or routines may lead to errors. On the contrary, particular attention to the concrete characteristics of each situation is required[7].

Medicine (which is the activity most studied by functionalists and interactionists) is (also) the prudential activity *par excellence*, and often taken as an example by Aristotle and his main commentators (ibid.; Ricoeur 2001), as it deals with the individuality of patients and the complexity of the human body and psyche. The uncertainty with regard to the diagnosis and the efficiency of treatments illustrates the general features of prudential activities very well. However, many other activities also encompass a significant prudential dimension. Other health care professions, those relating to social work, architecture, design, industrial project management, teaching, research, justice, policing, diplomacy, the conduct of war and more generally government and finance, are also activities where practical wisdom is outstandingly useful – or at least could be, because it is conspicuously lacking in some of them, for example, in finance (Stiglitz 2003). While our intent is not to list them all, this enumeration does let us appreciate that these activities are ones that face one or more of the following four sources of uncertainty: the complexity and singularity of people (for caregivers, educators, social workers or even police officers), social scenarios (police officers, magistrates, researchers, diplomats) or complex objects (architects, researchers, project managers in industry), as well as the impossibility of directly knowing past events that could determine the course of

7 For a more general discussion of the concept of practical wisdom or prudence, their uses and the challenges associated with them, please refer to Champy (2022).

action to be taken (police officers, magistrates), where once again, complexity prevents us from reaching certainty about the past based on what is known in the present.

In all these activities, prudent action is fragile for several reasons. While professional action based on scientific knowledge could invoke the strength of scientific certainty, practical wisdom is required when this knowledge leaves room for the subsistence of irreducible uncertainties. This is why prudent action is conjectural. Acting in a situation of uncertainty, the prudent person must sometimes rely on gambles which – even if counterbalanced by the experience of more or less similar cases – can turn out to be unfortunate. This requires knowing how to take the risk of making a mistake, and if necessary, showing boldness[8]. In addition, practical wisdom is deliberative, and these deliberations should often concern not only the activity's means, but also its ends. Sometimes, the complexity of the cases treated prevents each of the activity's ends from being satisfied, no matter how desirable these may be. For example, in medicine, we have to maximize the chances of a cure, while limiting the inconveniences and the risks of treatments. This gives prudent action a political dimension[9]. The conjectural and political dimensions of practical wisdom prevent us from finding all the answers to difficulties exclusively in the technique alone. They nuance these responses with an irreducible element of subjectivity, in contrast with the objectivity professed by the representatives of professions and by functionalists. The following will show the importance of this point.

The members of professions involving prudential practices are more exposed than others to having their decisions contested, not because they are bad professionals, but because they face complex and uncertain scenarios[10]. The public may then be tempted to withdraw their trust and turn to competitors, hence the particular vulnerability of prudential activities in terms of market competition. Furthermore, the arrival of poorly trained and inexperienced non-professionals on these markets would be all the more risky, given the fact that the cases treated are complex. This all explains why the public authorities have conferred protection against competition on certain activities, in exchange for guarantees regarding the

8 This remark is particularly important for readers working in Latin languages where the translation of *phronesis* as prudence or prudencia adds a connotation of pusillanimity, which is highly likely to produce a misinterpretation.

9 Research on palliative care has shown that, when confronted with curative routines which are difficult to justify, debates on professional practices become politicized (Castra 2003). This is the first distinctive feature of professions involving prudential practices (see footnote 3).

10 However, any professional activity also has its share of simple cases, which avoid the difficulties discussed here.

activity's training and control. Furthermore, not being able to sustain objectivity throughout their most difficult decisions makes professionals vulnerable to claims and interference from the non-professionals in organizations. Thus, the need for practical wisdom also accounts for the protection of professionals against external interference with their work[11]. The fragilities due to the high density of practical wisdom in certain activities therefore constitute a coherent explanation for the protection conferred within the framework of professionalization processes.

A second argument in support of this theory of professionalization is that activities protected from market competition and interference in organizations are activities requiring a significant amount of practical wisdom[12]. Conversely, certain activities with a high density of practical wisdom are not protected or are poorly protected, as in the case of social work[13]. However, the regularity is no less strong. It is perhaps even more so if we consider gradations. Even if the modalities differ from one country to another, medicine, the prudential activity *par excellence*, is also the most protected in all countries. Legal professions are quite comparable everywhere; architecture is protected in almost all industrialized countries, but less strongly. The question was even raised in the case of social work. Although a professionalization process was launched, it was unsuccessful (this aspect will be discussed later). On the other hand, many activities are not protected anywhere: blue-collar occupations; artisanal and commercial activities. The similarity of results produced within different national frameworks is sufficiently clear to show the need for having them addressed by a theory of professionalization.

The demonstration developed here also includes a third argument. In the period when members of certain activities struggled to obtain a professional status, the conception of professionalism underlying those struggles gave ample room to the idea of practical wisdom. Thus, at the turn of the 19th and 20th centuries, the descriptions of professional work in treatises and discourses borrowed heavily from

11 Similarities to several themes of Abbott's theory appear here: the idea of articulation between the abstract and the concrete; the idea of uncertainty about what to do; the idea that the objectivity of responses to a problem is more of a horizon than a reality; and finally, the idea of inference.

12 "Closed markets" (Paradeise 1984), which only protect against competition, fall under another theoretical explanation.

13 Further research would be useful to shed light on the reasons why prudential activities are more or less protected. The hints appearing in view of the historical material and the theories available relate to the density of practical wisdom, the length of inference chains, the relationships with other professions and the capacity to mask the political dimension of the work which is consubstantial with said activities but which, when perceived, can encourage interference from those in power.

the features of prudential activity. Let us consider the examples of medicine and social work.

1.3.2. *The professionalization of American medicine*

One of the criticisms formulated with regard to functionalism is that, especially in the United States, the profession's recognition and the reorganization of medical studies preceded the advent of medical practice, largely based on the application of scientific knowledge. The historical fresco that Paul Starr dedicated to American medicine between the mid-18th and the end of the 20th century confirms this observation: the protection that medicine began to obtain at the end of the 19th century followed an *aggiornamento* of the activity which began with a movement toward the concrete and a great distrust of the metaphysical abstraction at the beginning of the century (Starr 1982, pp. 54–56). After the first decades of the century, greater emphasis was increasingly placed on clinical observation. In addition, this achievement of recognition was accompanied by therapeutic skepticism, which lasted until the end of the century. Modestly, medicine then became more interested in the social conditions of illness, which translated into a more holistic and less technical approach, as well as a more preventive than curative perspective. Starr also emphasized the importance given to experience and judgment in the representation of doctors. While medicine was discredited in the first half of the 19th century, the reconquest of authority and recognition followed a general attitude strongly evoking the characteristics of practical wisdom.

Two institutions played a prominent role in the educational innovations of the end of the 19th century: Johns Hopkins University and Oxford University, where two personalities had considerable influence: William Welch and William Osler. A novelty in the courses offered lay in the rapprochement with science. In order to break away from training based on "didactic lectures" (ibid., p. 114), universities equipped themselves with laboratories and scientific research became part of the curriculum for future doctors. However, the prevailing conception of medical work continued to give a significant role to empirics and experience. In 1893, a reform in the training at Johns Hopkins gave equal importance to scientific research and clinical instruction. Of the four years of training, two were devoted to on-the-job learning. Starr particularly insisted on this point: "though Hopkins accentuated science, it did not stand for a narrowly technical vision of medicine; this was the secret of its special éclat" (ibid., p. 116). Furthermore, Welch and Osler were both committed to ensuring that future doctors had a solid general culture, which should guide their judgment. This aspect highlights their awareness of the non-technical dimension of the activity. By insisting on experience and general knowledge, and even if they did not use the concept, Welch and Osler offered a portrayal of the prudent man.

1.3.3. *Medicine and doctors according to Osler*

A teacher at Johns Hopkins and then dean of the medical school at Oxford from 1905, Osler gave numerous addresses (to medical students, nurses and practitioners of medicine) on his conception of medicine[14], which served as a reference not only in medicine, but also for other activities. The idea of practical wisdom permeated those addresses. The latter certainly included themes that the functionalist theory of professionalization had highlighted and even taken up at face value, such as "disinterestedness": Osler presented the medical career as a vocation[15], and balanced this life of sacrifice with the symbolic rewards associated with devotion to your neighbor. Likewise, science played a large part in his addresses, the clinician having to rely on the knowledge of three disciplines: chemistry, anatomy and physiology. At Johns Hopkins and then at Oxford, Osler effectively ensured that research laboratories in those disciplines were set up at the hospital and for training. He also insisted on the idea that science is not always enough to know what to do: "Is there then no science in medicine? Yes, but in parts only" (ibid., p. 36). "Wisdom" is required to make the right decisions and best care for the sick, several themes linking this conception of wisdom with the philosophical concept[16].

Osler emphasized the uncertainty inherent in medical work. In 1889, he warned students at the University of Pennsylvania:

> a distressing feature in the life which you are about to enter […] is the uncertainty which pertains not alone to our science and art, but to the very hopes and fears which make us men. In seeking absolute truth we aim at the unattainable, and must be content with finding broken portions (p. 7).

The processes of disease are so complex that it is excessively difficult to search out the laws which control them (p. 128). Failures are therefore inevitable despite scientific progress (p. 8). Osler also addressed the difficult tensions to be overcome between competing goals of this multidimensional work, of which cure, education and prevention are the essential dimensions (p. 125). We understand that medicine is "one of the most difficult arts in the world to acquire" (pp. 23–24).

---

14 These addresses were compiled during his lifetime. I refer here to the 2nd edition, from 1914: https://archive.org/details/2aequanimitaswit00osleuoft/page/12/mode/2up?ref=ol&view=theater.

15 The doctor strives "ten years for bread, ten years for bread and butter, and twenty years for cakes and ale" (ibid., p. 142).

16 Osler used the terms *wise* and *wisdom* on several occasions.

In front of the students, Osler extended the description of the activity with the presentation of his ideas on training and insisted as much on the moral qualities required for the doctor to face work-related difficulties as on the scientific knowledge required: equanimity, imperturbability, presence of mind, sound judgment, ability to make decisions, ability to control emotions and courage (pp. 132–133). Indeed, these qualities are those that philosophers attribute to prudent people (Aubenque 1963). In Osler's addresses, difficulty, fallibility and humility clearly go hand in hand (p. 39): in the face of uncertainty, doctors must be aware of their own limitations and know how to learn from their mistakes (p. 40). Training is only longer and more difficult as the acquisition of abstract knowledge does not suffice in itself. It is a long process along which many experiences accumulate. The quality of the doctor depends as much (or even more) on a dearly acquired culture as on knowledge which can be formalized and systematically transmitted[17]. Osler therefore valued experience highly, upon which the ability to make relevant judgments is based. He spoke of the need for broad, thorough and prolonged clinical training and the importance of placing the student in close contact with the patient, diversifying the experiences encountered in training as much as possible (p. 145). Osler played a large role in the establishment and diffusion of the internship, which is emblematic of non-systematic training, by osmosis. However, a crucial quality for learning from experience (i.e. from success and failure) is modesty (pp. 130–131), a personal quality that cannot be transmitted. According to Osler, wisdom dies with the doctor (p. 140).

1.3.4. *The "social case work" according to Mary Richmond*[18]

Osler's position is of even more historical interest as other professions have been inspired by his way of thinking on practice and training based on the requirement for adaptation to reality and the difficulties this entails. In response to a conference by Flexner, which rejected that social work had the qualities required to access the status of a profession, Mary Richmond published several works aimed at showing why this work is on the contrary highly qualified and can only be accomplished by professionals. One of these works is devoted to social diagnosis (Richmond 1917). We will focus here on another, more complete publication, because it is a later work and covers broader themes: *Social Case Work*, published in 1922. The title properly sums up its central argument: social work always deals with unique cases.

17 "The kindly fruits of the earth are easily grown; the finer fruits of the mind are of slower development and require prolonged cultivation" (p. 29).

18 I owe to Lilian Gravière the idea of bringing together the writings of Richmond and the idea of practical wisdom. On his own work, see, for example, Gravière (2013).

Passages devoted to the theme of professionalization show that the aim is to give social work access to the status of a profession. To do this, the book systematically addresses Richmond's conception of social work; however, the presentation, albeit not systematic, reflects a conception of transmission where "a record [of former cases] [...] can be unexcelled material for training other case workers" (p. 28): the presentation is built upon the examination of six cases in which individuals or families in difficulty received support. Inspired by these cases, Richmond develops themes similar to those of Osler. The idea of complexity comes up regularly to qualify both society and people, who must be assisted in finding a sense of their place and reintegration in society (p. 134). Richmond emphasizes the uncertainty of how this process would unfold (p. 106). She repeatedly mentions the need to individualize work[19], to approach each person as wholly as possible, according to a holistic method combining economic, educational, health, psychological, social and cultural considerations (pp. 98–99) with deeper insights (p. 106), taking into consideration both social resources and the negative influences that the person's environment may contain, and being quite prudent[20] to pay attention to details and keep them in mind. This attention to detail, which excludes taking refuge in abstraction, is also the condition for a broader view: "the professional worker who, in any field, has imagination enough to deal effectively with concrete things, to scrutinize them and 'put them together without abstraction', is also likely to be the one who can be trusted to see their larger relations" (p. 243).

Social workers must be wary of anything that standardizes or mechanizes their activity: by visiting a family at any time, they avoid routines that are prejudicial to a proper understanding of the person and their environment (p. 75); they must avoid reducing people to categories (p. 155). The work takes time (pp. 125, 142), and carrying it out in depth requires strength (p. 155), especially since the practical working conditions do not help (pp. 217, 252). Thorough as it may be though, this work cannot remove all the uncertainties inherent in the support process[21]: the accurate and definitive diagnosis of the person's problems can only be made at the end of treatment (p. 105).

Richmond's portrait of good social workers agrees with Osler's portrait of doctors on the qualities usually noted in prudent persons: humility, patience, a sense of observation, discernment, polyvalence, firmness and the ability to decide[22].

19 It is a real mantra in her presentation (see pp. 22, 34, 42, 92, 98–99, 150).

20 "Careful to keep in mind" (p. 47). As with Osler, any precise translation of *phronesis* is absent. However, "wise" also comes up several times (pp. 24, 42, 167).

21 The expression "deeper insights" (p. 105), almost an oxymoron, reminds us that deep work only allows us to gather insights, as opposed to an exhaustive analysis.

22 See pp. 24, 75, 80, 86, 90, 125, 143, 163.

Although these qualities are partly moral, they can be cultivated through experience. Thanks to them, the professional is better equipped than any other worker to gather the information required regarding a person and their environment, and then make an informed (i.e. personalized) decision.

1.3.5. *Professionalization theories and historical facts*

The attempt to obtain recognition of social work as a profession failed[23], which showed that working on singular and complex cases did not suffice. However, Richmond's work allows us to describe the features of professionalism as something very different from what had hitherto drawn the attention of sociological theories. Like Osler's addresses, it revealed that workers seeking to have their activity recognized as a profession placed less emphasis on scientific knowledge than on the difficulties and strong uncertainties their activity is confronted with, as well as on the need for anchoring work in concrete reality. The examples in Richmond's book illustrate even better than Osler's more general addresses the links between the impossibility of completely knowing a person and the relationships the person establishes with their environment, on the one hand, and the uncertainties that result from this, the difficulty of the decisions to be made, the inevitable conjectures and the boldness and discernment that the social worker must demonstrate, on the other hand. At the end of the 19th century and the beginning of the 20th, the members of the occupations seeking to access the status of a profession held a conception of professionalism that allows us to speak of *prudential professionalism*. Even if insufficient, the high density of practical wisdom is a reason for protection. This thesis avoids both the functionalist error overestimating the role of science, and the Beckerian aporia of explanation by contingency.

A perusal of the role played by case work in Richmond's research invites a more general comment: *case work* is present in other activities. We have seen it at work in medicine, with Osler's comments on teaching at the patient's bedside. Its source is found in law, with judgment commentaries and jurisprudence studies: casuistry is a discipline in itself. A variation of it is training games[24] and simulations. In all these

23 This is in fact only partial failure. The difference with recognized professions is that protection against competition is organized within the framework of employment, subjecting workers to strong hierarchical pressure. Social work is also sometimes referred to as a semi-profession. For a broad overview of social work activities, see Kuehni (2019).

24 Close but different from working on real cases, training games and simulations appeal to fictitious situations to learn the profession. In a thesis devoted to these, Léo Touzet shows the role of training exercises for war and their introduction by Vital Roux into management training, at the beginning of the 19th century (Touzet 2016).

areas, application is a skill in itself, because framing concrete cases against general rules requires discernment. As Schön has shown, this explains training modalities which greatly differ from didactic teaching based on abstract knowledge (Schön 1994). However, he also showed that this conception of professionalism declined in the 20th century to the benefit of the idea of "applied science"[25], with consequences that the idea of practical wisdom illuminates.

1.4. Objectivity, professionalism metamorphoses and vulnerabilities

1.4.1. *The end of the golden age for professions*

Unless it helped us to understand the present, a new professionalization theory would only be a scholastic exercise. As a matter of fact, this theory sheds light on a series of recent developments closely studied by the sociologists of professions and political scientists under various names: challenges to professionality (Ravon and Vidal-Naquet 2016), a decline in autonomy (Bezes et al. 2011), deprofessionalization[26], a decline in occupational professionalism in favor of organizational professionalism (Fournier 1999; Evetts 2003) and a degradation of the "service relationship" to "service provision" (Demailly 1998). These themes on developments share similarities, as evidenced by common topics: the role frequently attributed to new public management and more generally to managers, the primacy of economic goals over social considerations, the deterioration of working conditions, the feeling among professionals that it is increasingly difficult to conduct their work as they deem appropriate, or even the tension between an injunction to autonomy and resources and working conditions not conducive to the peaceful exercise of said autonomy[27]. These similarities lead us to wonder to what extent these labels designate substantially different processes or refer to scattered attempts to thematize similar (or even indistinct) processes. Here, the number of pages only allows us to respond by preterition. I would only like to show that these similarities or proximities stand out, and all the more so if we are interested in the prudential dimension of the activities concerned. The relationship with practical wisdom is also

25 It thus appears that functionalism, a good product of its time, came from the importation of this idea of applied science into sociology.

26 See the "deprofessionalization" dossier from the journal *Recherche et formation*, no. 72, 2013, and its discussion by Champy (2014).

27 On these injunctions to autonomy, the difficulties in responding to them and the suffering that this causes, please refer to the chapter by Arnaud Mias hereafter. These difficulties are not specific to the world of professions or work: they became a general feature of modern societies by the end of the 20th century. On this theme, see, in particular, Ehrenberg (2000) and Castel (1995).

explicit in several authors, notably regarding social work (Ravon and Vidal-Naquet 2016; Kuehni 2019), care activities (Pierru 2013) and architecture (Champy 2011).

Building on these works and on others which have not yet established this link, we will defend a thesis across four points. (1) One of the main challenges of these developments is the greater or lesser possibility left to professionals to conduct their work in a prudential manner. (2) Without constituting sufficient explanations for these developments, new public management and organizational professionalism strongly contribute to them, and their strength is closely linked to the general decline in the idea of practical wisdom. Bearing a conception of rationality from which the requirements of practical wisdom are absent, in contexts marked by strong contingency, organizational professionalism is engaged in this movement which encourages less practical wisdom. By implementing standards and systems which can be explained by this conception of rationality, new public management has an impact on the work of members exercising prudential activities. (3) These cultural and managerial developments subject professionals to challenges to professionality, which act as additional sources of complexity or uncertainty to their work. (4) The concept of practical wisdom helps us to see that the effects on the quality of services are a real and broader challenge than the effects on autonomy, because they concern the recipients of the work, not only the professionals.

To defend these theses, we will first study the relationships between organizational professionalism and new public management with practical wisdom. Then we will show how an inaccessible objectivity goal is imposed regarding professional practices. Finally, we will discuss the responses that professionals provide to the challenges these developments constitute.

1.4.2. *Organizational professionalism, new public management and practical wisdom*

In Fournier (1999) and Evetts (2003), the expression "occupational professionalism" designates a conception of professionalism which underpinned professionalization processes at their beginnings. This professionalism goes hand in hand with a demand for autonomy presented by professionals as essential to the quality of the service provided. Having emerged more recently, organizational professionalism diverges from the values of the former. It makes reference to the interests of the recipients of professional services to impose the following assessment criteria for gauging the quality of work: the predictability of the work process and its results, the conformity of the latter to quality standards, the respect of predefined deadlines for fulfilling the task and increasing the productivity of this task. Not only are these criteria that now define professionalism dissociated from the profession, but the values of organizational professionalism are unattainable in situations of irreducible uncertainty where practical wisdom is required.

The idea that work is too variable and should be more objective plays an important role in the attempts to make it more efficient: management thus bases its actions on a criticism of professions close to that of Freidson's early works (see above). However, the objectivity of a response to a problem would require the removal of all uncertainty. The concept of practical wisdom helps us to perceive that, as long as uncertainties remain, the choice of the most relevant actions can only be conjectural, and therefore subjective. While experience helps in choosing the most pertinent action[28], any risk of error cannot be completely ruled out. Likewise, the predictability of the time devoted to a piece of work is only possible for simple cases: in the face of complex and uncertain situations, time contains an element of unpredictability. This is why the demand for performance is based on a denial of the complexity of certain cases or certain tasks, and professionals who nonetheless try to satisfy it risk not taking such complexity sufficiently into account in their work. Research by Nicolas Belorgey on hospital emergency services showed that the injunction to reduce the waiting time of patients who spontaneously showed up at the medical service resulted in an increase in the rate of return to the hospital in the subsequent hours or days (Belorgey 2010). Working faster, caregivers make more decisions that do not take into account the uncertainties and possible dangers regarding the rest of the patients' health process sufficiently. Even if the requirements of organizational professionalism seem to be common sense (is not it better to be objective and efficient at work?), they can only be taken literally at the cost of negating the complexity of work in prudential activities.

Organizational professionalism is not, however, a creation totally foreign to professions, imposed on them from the outside to combat and supplant occupational professionalism (deemed corporatist). On the contrary, precisely because its values seem self-evident, they were already present in occupational professionalism. The search for the most objective possible answer explains the place given to science in the professionalization work carried out by Osler for medicine and, to a lesser extent, by Richmond for social work[29]. However, this interest in science was counterbalanced by an awareness of its limitations and work-related difficulties. Furthermore, the claim to objectivity responded to a social demand. In his history of American medicine, Starr narrates how, in the first half of the 19th century, doctors

28 In no way are all answers equivalent, and one of the interests of the concept of practical wisdom is that it offers a way out from the opposition between rationalism as a narrow conception of rationality and radical relativism (Champy 2022).

29 Mentioned in *Social Case Work*, science plays a slightly more important role in *Social Diagnosis* (Richmond 1917).

had identified this demand and learned to respond to it fallaciously, by presenting themselves to their patients as more self-assured than they actually were. Professionalism then came down to knowing how to exude an appearance of objectivity and certainty, even if this meant having to lie to customers[30].

Certainly, in the second half of the 19th century, the *aggiornamento* of medicine was driven by more modest and more realistic positions, which acknowledged the prudential dimension of the activity, to the point of making it the cornerstone of professional discourse. However, as medical progress became more substantial, the demand for objectivity and certainty could only increase, and it became ever more difficult to refer to uncertainties and difficulties to avoid having to provide answers. Expectations that are increasingly difficult to meet are aimed at doctors, and more broadly at members of professions involving prudential practices[31]. The conception of professionalism that is gradually establishing itself increasingly requires professionals to conform, thereby reducing the space left to the demands of practical wisdom. It is therefore understandable that sociologists ignoring authors like Osler and Richmond specifically insisted on the theme of objectivity in professional rhetoric, overlooking the fact that this was marginal at the beginning of the history of the professions (Parsons 1939, pp. 56–58; Paradeise 1985).

New public management thrives on this soil which is barely conducive to practical wisdom: culturally, its values are those of organizational professionalism, but it adds an important dimension to the developments already described. While practical wisdom presupposes a strong anchoring of action in a concrete reality, an essential characteristic of new public management is that it promotes the remote control of the activity of profession members (Evetts 2003), by poorly connected managers in the real areas where the professionals perform[32]. Through this remote control, new public management goes against what representatives of professions

---

30 Further in the past, this observation can be compared to the plays of Molière, in which doctors spoke Latin just to give the impression they were tapping into mysterious knowledge, when they were actually being inconsistent.

31 These excessive expectations of what professionals can do have been brought to light by Beck's work on the growing demand for security in an uncertain world (1986). More recently, works by Marc Loriol on fatigue (Loriol 1999), and Sébastien Dalgalarrondo and Boris Hauray on aging (Dalgalarrondo and Hauray 2015) study the construction of new medical problems at the crossroads of rising customer expectations and medical responses, which are sometimes less well-founded than doctors may try to make us believe.

32 This also happens when managers have been trained within the profession, as is evident in the distinction made by Freidson between practitioners, teachers and managers in the same profession (three categories whose experiences and values can only drift apart over time).

like Osler and Richmond had clearly grasped: that the quality of work largely depends on its anchoring in concrete reality.

To assume this remote control, new public management has recourse to a set of tools properly studied by sociologists: indicators, rules and procedures for evaluation and accountability. Let us go back to the example of the hospital emergency department studied by Nicolas Belorgey. The time for which patients wait for a medical response is interpreted as information regarding work quality. Belorgey shows the perverse effects of its use. The error is to turn this indication – which is only one among many required to assess work – into an exclusive indicator whose value is unquestioningly considered with regard to what it can reflect/produce; in other words, with no practical wisdom. These indicators producing results contrary to the ones expected from their implementation therefore constitute a recurring problem for organizations (Beauvallet 2009). Broadly speaking, the link between indicators and a lack of practical wisdom has been clarified by Porter's work, who showed that the recourse to numbers is often based on an irrational faith in their relevance (Porter 1995). Practical wisdom, however, is not opposed to numbers, which are often essential to describing a situation, but practical wisdom does not allow their mechanical use and their fetishization. It presupposes a conscientious pondering of their deeper meaning and the concrete reality they supposedly mirror. The prudential use of numbers starts from the observation that they constitute schematizations of reality and must be taken as such, contemplating what assumptions their construction is based upon, to what uses they are adapted and what parts of reality they leave outside their scope. It is not the figure that poses a problem. It is the distance its use allows, contravening the prudential anchoring of action in concrete reality, and the feebly grounded certainties it generates.

Exactly the same happens with rules, another central tool in the arsenal of new public management. Rules can help in the prudential conduct of work: quality standards require certain dimensions of work to be effectively taken into account; highly coded legal procedures make it possible to avoid errors and injustices by protecting prudential forms of deliberation. However, this contribution to the construction of a work framework conducive to practical wisdom depends on three sets of rule characteristics: their degree of adaptation to the concrete reality, their greater or lesser degree of flexibility and their stability. In fact, new public management, and more generally remote management, presents significant shortcomings from each of these three perspectives. In a summary article on "organizational work", Marie-Anne Dujarier evokes the frequent failures of hierarchy in adapting rules (Dujarier 2006). Organizational work is defined as an effort which should continually adapt rules to the reality of the work they govern. It is based on negotiations and exchanges between successive levels, forming a pyramid going from the recipients of the service and the workers engaged in concrete work, up to the organization's management[33]. This work should enable the organization to adapt to tasks and the environment, limiting the time and extent of the inevitable mismatches between the organization's goals and concrete practices. Extensive research has shown that this work is no longer done, or only at the lower levels (ibid.). Higher levels of the hierarchy ignore (or feign ignorance of) the complexity of the lower levels, forcing the latter to shoulder the gap between what should be done (prescribed work) and what can actually be done. This is where the second characteristic of rules comes into play. These inadequacies would be easier to bear if professionals were authorized to free themselves from the rules when a case or situation clearly highlights the benefit of such transgression. However, that is not the case. On the contrary, surveys point out the obligation professionals often find themselves in to apply a rule they consider harmful, and the personal risks they would take otherwise. Professionals must learn to deal with both the complexity of the cases and situations they deal with, and the constraints of the rule system. In fact – and this is the third point – learning often has to be restarted, because these rules change at a sustained pace, which adds uncertainties to those consubstantial with prudential activities[34].

In short, the narrow conception of rationality underlying organizational professionalism and new public management ignores the demands of practical wisdom. As a result, the activities most affected by these "rationalist" logics should be the professions involving prudential practices. A look at the activities which have given rise to work or debate on this theme supports this thesis. After all, it is in the prudential professions that the observations of a deterioration in activity have been most documented. Countless works concern social work (Kuehni 2019), research (Bruno 2008), the judiciary field (Vauchez and Willemez 2007), architecture (Champy 2011), health care professions (Juven et al. 2019) or even the police (Lemaire 2016). As a matter of fact, less prudential activities also face difficulties due to the intensification of work or changes in their supervision, but as far as they are concerned, most challenges concern domination relationships within the framework of the organization or the company (Champy 2014). In prudential activities, on the other hand, the tasks to be carried out are directly affected at their

33 Marie-Anne Dujarier differentiates between the organizational work successively carried out by (1) boards of directors; (2) upper managers assisted by experts; (3) lower managers and their teams; (4) local operational teams; (5) interactions between the employee and the customer or service user; and (6) each worker during the performance of a task.

34 On these last two points, see the chapter by Arnaud Mias in this volume, the article by Dujarier or for an example relating to social work (Lascar 2016).

core and the meaning of work itself is at stake: two paradigms oppose one another. Since the 1990s, attempts to implement evidence-based practice (EBP), based on the evidence-based medicine (EBM) model, have accurately reflected the tensions between these opposing paradigms.

1.4.3. *The stubborn mirage of the scientificity of practices*

Associated by some with the triumph of rationality and criticized by others, EBM represented an important innovation in medical practice. Its development was based on a distrust of doctors' experience: relying on the latter leads to heterogeneous practices, some of which are necessarily less relevant than others (Bergeron et al. 2015, pp. 140–148). EBM aims to identify and promote best practices through three mechanisms: therapeutic trials used to objectify treatment efficiency, a meta-analysis of data collected across different trials, and best practices guidelines that make the results obtained available to caregivers. The goal is to reach greater objectivity. Bergeron et al. (2015) identified the limitations of EBM, which can in no way be reduced to a triumph of rationality. These limitations include the heterogeneity of best practices guidelines for the same pathology; situations where the application of guidelines would pose a risk; situations where the guidelines point to several options; and political uses. According to the authors, EBM shifts uncertainty rather than reducing it, which leaves an irreducible place for individual judgment in decisions. Now, this illustrates the profound nature of medicine as a prudential activity: despite having divergent principles, factual medicine and practical wisdom are not opposed in practice. On the contrary, while EBM does not abolish prudential judgment, practical wisdom also involves relying on existing relevant data to best adapt treatments to patients' pathologies. However, best practices guidelines must be produced and used with practical wisdom. The challenge lies in the articulation of formalization and reflection, which seems very successful in medicine, according to the research and the broad consensus to which EBM was recently subject (even if it does not reach the pure and complete rationality some attribute to it, since this limitless rationality is inaccessible in a situation of uncertainty).

The awareness that factual medicine does not achieve its objectivity ideal nor does it suspend all judgment has not hindered attempts to apply its logic with unfounded rationalist hopes to even more rebellious epistemological areas (less well anchored in science than medicine). Twenty years ago, Trinder and Reynolds (2000) showed the role played by managerial culture in the diffusion of this logic to areas like probation practices, social work, education and human resource management. What the authors showed at that moment still holds true at present. EBP devalues the experience of professionals and aims to reduce their autonomy. But at least two things are missing for it to be as relevant as in medicine: reliable, easily measurable data, which are recognized as relevant by segments of the professions whose judgment is considered as relevant because of both their size and the positions of their representatives; sufficiently clear and shared goals for action. Correspondingly, the actors who promote EBP use it far from the methodological canons of EBM, and even in more various ways. For example, double-blind trials are generally absent from protocol development. However, these differences are still blind spots, and such forms of EBP are not subject to the type of evaluation they deem essential for professional practices. Hence, the idea that EBP only promotes a dangerous illusion of objectivity:

> a particular version of science is a blunt or brutal instrument for dealing with the complexity of human behavior, emotions and relationships. Extending this further it is possible that the apparent cleanliness and transparency of science may go beyond failing to do justice to complexity, and may even simplify and distort the picture that is seen, with significant consequences for practice (ibid., p. 151).

Finally, despite these long-standing criticisms, attempts to base the work on conclusive data do not falter for education (Saussez and Lessard 2009; Larose et al. 2011), social work (Kuehni 2019) or psychology (Durieux et al. 2017). On the contrary, they are progressing, including in the analysis of public policies, through evidence-based policy making (EBPM), for which the same criticisms against "overly technicist and rationalist perspectives" are resurfacing (Castel 2018).

EBP illustrates what is at stake in the attempts from management to control professions involving prudential practices: a denial of complexity, uncertainties and the irreplaceable role of experience in coping with them. However, the anti-prudential logic underlying both the blindness to EBM's limitations and the efforts to impose it appears even more clearly when considering examples of the same logic from outside the field of professional work. Attempts to extend the principles of EBM to activities less anchored in science than medicine are actually based on a mythology omnipresent in society and close to that which Porter described regarding the use of numbers (Porter 1995). From the fetishization of GDP to scientific criticisms of the precautionary principle via the practices of financial capitalism, the denial of uncertainties, false objectivity and reckless risk-taking are widespread and closely linked (Champy 2022, pp. 185–206). All these myths, which presuppose an uncritical faith in the possibility of submitting everything to supposedly objective methods, show that the evolution of professions pertains to a more general social fact: the weak recognition of practical wisdom in industrial societies at the end of the 20th and the beginning of the 21st century.

1.4.4. *Mobilizations and challenges to professionality*

Management offensives affecting the prudential content of work take other forms, too. Even in architectural contests, where architects are expected to produce original projects, the theme of the quantified "objectivity" of responses appeared in the deliberations of juries about 25 years ago (Champy 2011). Furthermore, the injunction to performance also produces nefarious effects illustrated by the consequences on the uses of waiting times in emergency rooms, as studied by Nicolas Belorgey. In matters of justice, the implementation of "real-time processing" in district courts reduces the time spent examining files and deliberating on a decision, producing a form of decision mechanization (Bastard and Mouhanna 2007). Finally, while the importance of the relationship between the police and the field was sufficiently recognized for management to step aside in favor of the base, according to a principle of "hierarchical inversion" (Monjardet 1996), the managerial reforms at the beginning of the 21st century have strengthened remote control, leading to a "hierarchical redistribution of work" (Lemaire 2016).

These paradigmatic tensions are experienced in diverse ways by the professionals engaged in work. Obstacles to practical wisdom pose a problem for many of them. Some have come together in protest rallies, as exemplified by *l'Appel des appels*: academics, teachers, caregivers, social workers, artists and even magistrates joined together to analyze and denounce the difficulties caused in their work by recent reforms, such as the French law on the Liberties and Responsibilities of Universities, and that on Hospitals, Patients, Health, Territories, which affects hospitals[35] (Gori et al. 2009). It is remarkable that the activities represented in this collective are prudential activities[36], even if all the activities involving prudential practices are not represented (it is impossible to draw up a list for all of them). For example, the police are absent, but police action campaigns have existed in other forms. These various demonstrations reveal a new mode of professional socialization, affecting activities, such as medicine and the police, which used to be reluctant to do so. They are bringing to life collectives and prompting exchanges that contribute to creation of people's relationship to their occupation.

Even more interesting are the reactions in the workplace, where the different conceptions of professionalism come into tension. Professionals seduced by the

35 The French laws on the Liberties and Responsibilities of Universities (2007) and on Hospital, Patient, Health, Territories (2009) aim to rationalize spending and put organizations at the service of their users, at the cost of strengthening managerial logics such as benchmarking.

36 Or occupations evolving in a framework where practical wisdom is crucial, like the nursing assistants opposed to the HPST law.

sirens of objectivity embrace managerial logics and take part in furthering these: the role of some in the dissemination of various forms of EBP attests to this. Others fight this logic and take initiatives to restore spaces for practical wisdom. In the field of medicine, for example, greater inter-professionality helps in promoting overviews of cases; during training, multiple choice questionnaires based on the idea that a single answer exists are replaced by script concordance tests, as implemented in Montreal (making it possible to evaluate clinical reasoning); as done by the journal *Prescrire*, reasoning biases induced by conflicts of interest with the pharmaceutical industry are fought against; or even some doctors giving a place to the patient's singular history, as in the narrative medicine model developed at Harvard. As happened with Richmond and Osler at the turn of the 19th and 20th centuries, these actors are generally ignorant of the concept of practical wisdom, but call upon its main idea: they are "practical wisdom entrepreneurs" (Champy 2018).

The most common attitude is to pursue work while trying to limit the consequences of the mismatch between what professionals think they should do and what they can actually do. To study difficulties in adaptation, Bertrand Ravon and Pierre Vidal-Naquet talk about "tests of professionality" (Ravon and Vidal-Naquet 2016, 2018). This notion originated from the observation that "there are numerous examples of nurses or caregivers who, confronted with grueling tasks which have to be carried out under the pressure of efficiency or urgency, consider their work impossible and unbearable" (ibid., p. 75). Establishing an explicit link with the requirements of practical wisdom, the authors propose:

> a definition of the challenge to professionality centered on the disorders experienced within the action itself, when professionals report great perplexity in the face of the uncertainty of the situations encountered, because they are no longer able to discern exactly what is appropriate to do (Ravon and Vidal-Naquet 2016, p. 74).

The same observation could be made about researchers (Bruno 2008) or legal professionals confronted with mass litigation and justice reforms (Vauchez and Willemez 2007). Ravon and Vidal-Naquet clearly show that these challenges are largely due to the inadequacy of the work organization and the rules framing the complexity of such work, thus joining the aforementioned studies on "organization work". To be able to carry out their work in an acceptable way for them, professionals equip themselves with a self-mandate that deviates from prescribed work. "Torn between contradictory standards [...], workers are led to redefine what should be done from one day to the next, based on the circumstances and beliefs of the moment" (Ravon and Vidal-Naquet 2018, p. 79). However, a major difficulty is that "the legitimacy of self-mandate is never acquired" (ibid., p. 81). The solution to the doubts and suffering caused by these situations may lie in avoidance practices and disengagement from work. On the contrary, in other cases, professionals adapt themselves through increased reflexivity and inventiveness, which can contribute to renewing the significance of work. Challenges to professionality are modifying the exercise of practical wisdom. In addition to the sources of complexity and uncertainty which are consubstantial with prudential activities, there are also sources of complexity and uncertainty inherent in the contexts where work takes place, and which could undoubtedly be avoidable, at least in part.

1.5. Conclusion

The philosophical concept of practical wisdom sheds light on the historical processes through which professions have accessed protections denied to other occupations, and thus answers a key question in the sociology of professions. Parsons found an answer in the mastery of scientific knowledge, but this answer is not consistent with the observation that science actually has a limited place in the history of professions and their practices. Arguing on the contingency of historical processes, interactionists have renounced any theoretical explanation. Their position does not stand up to the observation that different countries protect the same activities. Taking up the theoretical ambition of functionalism while integrating the contributions of interactionism, Abbott thematized the way in which activities obtain a formal protection in their jurisdiction; however, he failed to account for the establishment of protection. Furthermore, as the latter is likely to hinder the adaptation of the division of labor to efficiency requirements, to mention them seems to invalidate the important role that the theory recognizes in the efficiency. The explanation through practical wisdom avoids these difficulties altogether. It detaches protection from the scientific nature of work, it accounts for the regularity in protected activities across different countries, it clarifies the reasons for protecting certain activities, and it is supported by the massive presence of the idea of practical wisdom in professional discourses. However, the case of social work shows that certain professions involving prudential practices are poorly protected, if protected at all. New research could shed light on the conditions for a prudential activity to access professional status.

This way of revisiting professionalization theories sheds light on the present. The professions whose practical wisdom has justified protection are those which suffer the most from developments studied by researchers under various names. Moreover, the transition from protection to attacks in less than a century can easily be understood in light of the content of professionalization theories. The functionalist theme of science is a reflection of the broad success of the theory of applied science, which exposed protected activities to a demand for objectivity which was impossible to satisfy. Criticizing the lack of scientificity of work, the interactionists took up the aspiration of their time for the professionalization of everyone and for the lack of distinction between professions and occupations, despite Strauss's intuition regarding the very unequal complexity of work. The successive sociological impasses reflect those of their time. It is therefore not surprising that certain professionals have also played their part, first by putting objectivity at the heart of their rhetoric, then by supporting management in its quest for an inaccessible objectivity.

Prudentialist theory inevitably also has links with the trends of its time, but its aim is to open up reflection without taking sides regarding the challenges professionals face. While the sociology of professions continues to take a keen interest in the question of the autonomy of professionals, the idea of practical wisdom makes it clear that the adaptation of work to the cases and situations encountered constitutes a broader challenge, because it concerns all the recipients of work and not only professionals. In fact, autonomy is ambivalent in this respect: it seems to be required as the application of a principle of subsidiarity for the prudential anchoring of work in concrete reality, but it can be misused. To escape from this circle, the concept of practical wisdom invites a questioning of what promotes or hinders it in various contexts. Under this condition, this sociology can open up a horizon of common reflections for actors with divergent interests.

1.6. References

Abbott, A. (1988). *The System of Professions. An Essay on the Division of Expert Labour.* University of Chicago Press, Chicago/London.

Aubenque, P. (1963). *La prudence chez Aristote.* PUF, Paris.

Bastard, B. and Mouhanna, C. (2007). *Une justice dans l'urgence. Le traitement en temps réel des affaires pénales.* PUF, Paris.

Beauvallet, M. (2009). *Les stratégies absurdes. Comment faire pire en croyant faire mieux.* Le Seuil, Paris.

Beck, U. (1986). *Risk Societies. Towards a New Modernity.* Sage, London.

Becker, H.S. (1962). The nature of a profession. *Teachers College Record*, 63(10), 27–46.

Becker H.S., Geer B., Hughes E.C., Strauss A.L. (1961). *Boys in White. Student Culture in Medical School.* Transaction Publishers, New Brunwick/London.

Belorgey, N. (2010). *L'hôpital sous pression. Enquête sur le "nouveau management public".* La Découverte, Paris.

Bergeron, H., Castel, P., Hauray, B. (2015). Evidence-based medicine. In *Dictionnaire critique de l'expertise*, Henry, E., Gilbert, C., Jouzel, J.-N., Marichalar, P. (eds). Presses de Sciences Po, Paris.

Bezes, P., Demazière, D., Le Bianic, T., Paradeise, C., Normand, R., Benamouzig, D., Pierru, F., Evetts, J. (2011). New public management et professions dans l'Etat : au-delà des oppositions, quelles recompositions ? *Sociologie du travail*, 53(3), 293–348.

Broadie, S. (1991). *Ethics with Aristotle*. Oxford University Press, New York.

Bruno, I. (2008). *A vos marques®, prêts... cherchez ! La stratégie européenne de Lisbonne, vers un marché de la recherche*. Éditions du Croquant, Boissieux.

Castel, R. (1995). *Les métamorphoses de la question sociale. Une chronique du salariat*. Fayard, Paris.

Castel, P. (2018). Compte rendu : P. Cairney, the politics of evidence-based medicine. *Gouvernement & action publique*, 7(3), 123–126.

Castra, M. (2003). *Bien mourir. Sociologie des soins palliatifs*. PUF, Paris.

Champy, F. (2011). *Nouvelle théorie sociologique des professions*. PUF, Paris.

Champy, F. (2012). *La sociologie des professions*. PUF, Paris.

Champy, F. (2014). Faut-il parler de déprofessionnalisations au pluriel ? *Recherche et formation*, 75, 133–146.

Champy, F. (2018). The sociology of prudential activities: From collective commitment to social innovations. *Sociologia Problemas e Práticas*, 88, 79–94 [Online]. Available at: https://revistas.rcaap.pt/sociologiapp/article/view/14799.

Champy, F. (2022). The imperative of practical wisdom in the face of anthropocene emergencies: The case of climate change. In *Risks and the Anthropocene*, Rebotier, J. (ed.). Wesley, London.

Dalgalarrondo, S. and Hauray, B. (2015). Les économies de la promesse anti-âge. Le cas de la DHEA. *Sciences sociales et santé*, 33, 5–30. doi: 10.3917/sss.332.0005.

Demailly, L. (1998). Les métiers relationnels de service public : approche gestionnaire, approche politique. *Lien social et Politiques*, 40, 17–24.

Demailly, L. and De La Broise, P. (2009). Les enjeux de la déprofessionnalisation. Etudes de cas et pistes de travail. *Socio-logos*, 4 [Online]. Available at: http://socio-logos.revues.org/2305.

Dujarier, M.-A. (2006). La division du travail d'organisation dans les services. *Nouvelle revue de psychologie*, 1, 129–136.

Durieux, N., Étienne, A., Willems, S. (2017). Introduction à l'*evidence-based practice* en psychologie. *Le Journal des psychologues*, 345, 16–20. doi: 10.3917/jdp.345.0016.

Ehrenberg, A. (2000). *La fatigue d'être soi. Dépression et société*. Odile Jacob, Paris.

Evetts, J. (2003). The sociological analysis of professionalism: Occupational change in the modern world. *International Sociology*, 18(2), 395–415.

Flexner, A. (1915). Is social work a profession? *Research on Social Work Practice*, 11(2), 152–165.

Fournier, V. (1999). The appeal to 'professionalism' as a disciplinary mechanism. *The Sociological Review*, 47(2), 280–307.

Freidson, E. (1970). *La profession médicale*. Payot, Paris.

Freidson, E. (1986). *Professional Powers. A Study of the Institutionalization of Formal Knowledge*. Chicago University Press, Chicago.

Freidson, E. (1994). *Professionalism Reborn. Theory, Prophecy and Policy*. Oxford Polity Press, Oxford.

Freidson, E. (2001). *Professionalism. The Third Logic*. Oxford Polity Press, Oxford.

Gori, R., Cassin, B., Laval, C. (2009). *L'appel des appels. Pour une insurrection des consciences*. Fayard/Mille et une nuits, Paris.

Gravière, L. (2013). L'empirisme démocratique de Mary Richmond. *Vie Sociale*, 4, 99–113.

Héran, F. (2021). *Lettre aux professeurs sur la liberté d'expression*. La Découverte, Paris.

Hughes, E.C. (1984). *The Sociological Eye. Selected Papers*. Transaction Books, New Brunswick.

Juven, P.-A., Pierru, F., Vincent, F. (2019). *La casse du siècle. A propos des réformes de l'hôpital public*. Raisons d'Agir, Paris.

Kerleau, M. (1998). L'hétérogénéité des pratiques médicales, enjeu des politiques de maîtrise des dépenses de santé. *Sciences sociales et santé*, 16(4), 5–32.

Kuehni, M. (ed.) (2019). *Le travail social sous l'œil de la prudence*. Éditions Schwabe, Basel.

Larose, F., Couturier, Y., Bédard, J., Charette, S. (2011). Entre discipline et profession : la question des bonnes pratiques guidées par les résultats probants de la recherche en formation à l'enseignement. *Les Sciences de l'éducation - Pour l'Ère nouvelle*, 44(2), 31–48. doi: 10.3917/lsdle.442.0031.

Lascar, E. (2016). "Incasables". Une mise à l'épreuve de la protection de l'enfance. Dissertation, Institut Méditerranéen de Formation et de Recherche en Travail Social, Marseille.

Lemaire, É. (2016). Les usages de la spécialisation dans la police : les formes discrètes du management public policier. *Revue française de science politique*, 3(3–4), 461–482.

Loriol, M. (1999). *Le temps de la fatigue. La gestion sociale du mal-être au travail*. Anthropos, Paris.

Malatesta, M. (2006). *Professionisti e gentiluomini. Storia delle professioni nell'Europa contemporanea*. Einaudi, Turin.

Merton, R.K. (1957). Some preliminaries to a sociology of medical education. In *The Student Physician. Introductory Studies in the Sociology of Medical Education*, Merton, R.K., Reader, G.G., Kendall, P.L. (eds). Harvard University Press, Cambridge.

Monjardet, D. (1996). *Ce que fait la police. Sociologie de la force publique*. La Découverte, Paris.

Osler, W. (1905). *Aequanimatus and Other Addresses*. HK Lewis, London.

Paradeise, C. (1984). La marine marchande française : un marché du travail fermé ? *Revue française de sociologie*, 25(3), 352–375.

Paradeise, C. (1985). Rhétorique professionnelle et expertise. *Sociologie du travail*, 37(1), 17–31.

Parsons, T. (1939). The professions and social structure. In *Professionalization*, Vollmer, H.M. and Mills, D.L. (eds). Prentice-Hall, Englewood Cliffs, NJ.

Parsons, T. (1968). Professions. In *International Encyclopaedia of the Social Sciences*, Sills, D.L. (ed.). Macmillan, New York.

Pierru, F. (2013). Impératifs gestionnaires et *phronesis* médicale : esquisse sociologique d'un engagement éthique dans un grand hôpital parisien. *Quaderni*, 82, 67–82.

Porter, T.M. (1995). *Trust in Numbers: The Pursuit of Objectivity in Science and Public Life*. Princeton University Press, Princeton, NJ.

Ravon, B. and Vidal-Naquet, P. (2016). L'épreuve de professionnalité : de la dynamique d'usure à la dynamique réflexive. *Sociologies* [Online]. Available at: journals.openedition.org/sociologies/5363.

Ravon, B. and Vidal-Naquet, P. (2018). Les épreuves de professionnalité, entre auto-mandat et délibération collective. L'exemple du travail social. *Rhizome*, 67, 74–81.

Richmond, M. (1917). *Social Diagnosis*. Russell Sage Foundation, New York.

Richmond, M. (1922). *What is Social Casework: An Introductory Description*. Russell Sage Foundation, New York.

Ricœur, P. (2001). *Le Juste II*. Editions Esprit, Paris.

Saussez, F. and Lessard, C. (2009). Entre orthodoxie et pluralisme, les enjeux de l'éducation basée sur la preuve. *Revue française de pédagogie*, 168 [Online]. Available at: journals.openedition.org/rfp/1804.

Schön, D. (1994). *Le praticien réflexif. A la recherche du savoir caché dans l'agir professionnel*. Les Editions Logiques, Montreal.

Starr, P. (1982). *The Social Transformation of American Medicine*. Basic Books, New York.

Stiglitz, J.E. (2003). *The Roaring Nineties. Seeds of Destruction*. Allen Lane, London.

Strauss, A., Fagerhaug, S., Suczek, B., Wiener, C. (1985). *Social Organization of Medical Work*. Chicago University Press, Chicago.

Touzet, L. (2016). Jouer à faire des affaires. Une sociologie des business games. PhD Thesis, Université Toulouse Jean-Jaurès, Toulouse.

Trinder, L. and Reynolds, S. (eds) (2000). *Evidence-Based Practice. A Critical Appraisal*. Blackwell Publishing, London.

Vauchez, A. and Willemez, L. (eds) (2007). *La justice face à ses réformateurs (1980-2006). Entreprises de modernisation et logiques de résistance*. PUF, Paris.

Wilensky, H. (1964). The professionalization of everyone? *American Journal of Sociology*, 70(2), 137–159.

2

Professionalization: The Mystery of Boundaries

Didier DEMAZIÈRE
CSO, CNRS, Sciences Po, Paris, France

The notion of professionalization is widespread in all work environments, if not in society as a whole (Perkin 1988), and this does not go without the risk of diluting its meaning. If everything becomes professional, if everything is professionalizing or – to use a neologism – "professionalizable", then we can call into question the relevance of a concept that has become not only elastic or shapeless, but also unavoidable and omnipresent, when we wish to account for the transformations affecting métiers, production systems, market relations and the world of training. By spreading rapidly over the last decades in various fields of study such as the sociology of organizations (Demazière and Pélisse 2022, pp. 61–74), public action (Le Bianic and Vion 2008), administration and state (Bajard et al. 2018), political representatives (Michon and Ollion 2018) markets and customers (Dondeyne 2002) or education and training (Lang 1999), the notion of professionalization has been summoned to address diverse issues. However, these point more or less toward a central question, which can be formulated in terms of the consistency of the professional fact. By this expression, we must understand the modes and degrees of affirmation within organized (bureaucracies, markets, etc.) and regulated (by laws, quality standards, etc.) systems of workers practicing the same activity.

This perspective can be understood as the unfolding of the fundamental meaning of the concept of professionalization, as it circulates in its preferred field: the sociology of professions. In fact, it designates the processes of emergence, constitution and institutionalization of social entities aggregated around the exercise

Encyclopedia of Professionalization,
coordinated by Didier DEMAZIÈRE and Richard WITTORSKI.

of a particular occupational activity with the goal of protecting it, controlling its regulation, promoting valorization, or even conquering its monopoly. In a certain way, professionalization tells the stories of boundaries, whether this involves identifying particular professional activities, or building forms of protection to perpetuate them, or even constructing regulations to make them stand out as different from others. It designates the delineation or demarcation of boundaries between professionals and nonprofessionals, between more or less delimited professional groups, protected by their boundaries to a larger or lesser extent. The central challenge of professionalization lies in this "more or less", which qualifies the consistency, recognition, coalescence, identity, legitimacy of these groups, constituted around (more or less) specialized and identified activities in the division of labor.

It is this thread of boundaries that we will explore throughout the chapter, less to interrogate the delimitation of what a profession is than to provide information on what constitutes a profession: the very process of differentiation of work activities, their specialization, their affirmation and their recomposition. Whether we go through the analysis of work and concrete activities or we study political enterprises striving to conquer a professional territory, the goal is not to distinguish between categories of activities, but rather to inform the processes of professional categorization. How do activities emerge, how are they constituted, how do they differentiate themselves, how do they acquire recognition? How are boundaries asserted, moved, weakened, crossed, how do they disappear, etc.?

We will begin by showing that in its original sense the concept of professionalization is the operator making it possible to characterize the position of professional groups in relation to the status of profession (established, recognized), to index the distance to this reference figure along conditional criteria and to establish a tendentially watertight boundary between *professions* and *occupations*, a differentiation used in English-speaking countries. Then, we examine a conception of professionalization (more open and less centered on the culmination of the process), which conceives it as a collective action invariably distributed and engaging multiple stakeholders, which leads to focusing on the dynamics of the boundaries of professional groups. Professionalization is then conceived as work at the boundaries of any professional group, rather than as a generator of the boundaries which differentiate certain professional groups along the status scale. We will conclude by emphasizing that all professional groups, regardless of their organization and recognition, face the same challenge of asserting a place in the division of labor and that professionalization is a key to analyzing the recomposition of occupational activities, provided that it is conceived both as a general and selective process.

2.1. Professionalization as the selection of professions

A first chain of significations conveyed by the concept of professionalization points to the debates on how métiers and professional groups are characterized and granted a theoretical status. Professionalization is defined as a pathway, punctuated by obstacles to overcome. It is organized into multiple stages, leading to the attainment of a specific status: the status of *profession* as understood in the English language, where it denotes an established or institutionalized profession, holding the statute as such. The collective entities associated with the exercise of occupational activities can follow a path toward professionalization, along which they are more or less advanced and count on more or less resources to overcome pitfalls. Considering that not all of them manage to complete this entire path, professionalization is therefore a selective process: it is defined here as the selection of *professions* among a larger set of activities (among other *occupations*, in the English language sense). It is a marker of distinction, delineating a boundary between professions and other activities which are not recognized as such.

This boundary has a double signification, which reminds us of the two modes of performativity boundaries have (Gottmann 1980, pp. 53–58), matching the metaphors of the "door" and the "bridge" (Simmel 1988, pp. 159–166). On the one hand, the boundary dissociates, divides and selects. It includes and excludes, and in that case corresponds to the figure of a line or a door. On the other hand, the boundary lets pass, connects and links. It promotes circulation and acts like a zone or a bridge. Through its procedural dimension, the concept of professionalization is associated with this second logic, as a possible flow or path across this bridge, whose crossing could grant access to the status of a profession. But it also means that occupational activities are distributed among those who have passed to the other side, those who have crossed the door separating "real" professions and nonprofessions. Hence, we understand that this boundary is a major subject born with the emergence of the sub-field of sociology of work, whose specialization object is occupational activities and the division of labor into métiers, work specialties or expertise. In fact, it concentrates a series of debates on the way of designating the objects analyzed (professions or occupations, to use the two original categories, métiers, professional groups, expert work, etc.), the definition of the criteria for grounding such distinctions and the distribution of countless occupational activities on both sides of the boundary. It also gives rise to debates on the very relevance of the concept of boundary. We will first consider the question of professionalization as the access to a distinctive profession status. Following this, we will discuss the challenges related to drawing a demarcation line between certain professions and ordinary occupational activities. Finally, we will emphasize the risks of reducing professionalization to its outlets and neglecting the processes as such (an issue which arose with the diversification of research fields).

2.1.1. *Professionalization for all?*

The concept of professionalization has established itself as an essential tool for analyzing the world of labor, its use being profuse and widespread (Demazière et al. 2012). While it concentrates on the theoretical clashes between the classic trends in the sociology of professions from English-speaking countries (see Chapter 1 on this point), it has only recently spread in research on work and occupational activities in France. Notwithstanding this, it quickly became a concept dear to sociologists of work, as soon as they adopted the focus on métiers. It now runs through research devoted to the study of the most diverse occupational activities: physicians (Danet 2008), police officers (Pruvost 2007; Pichonnaz 2017), home helpers (Bonnet 2015, pp. 95–113), artistic intermediaries (Lizé and Naudier 2015, pp. 159–176), postmen (Salaün 2008), teachers (Guibert and Lazuech 2010), prison guards (Mallochet 2007; Combessie 2009), farmers (Nicolas 2018), etc., to mention only a few examples.

This work and many others have fueled questions about the identification of possible distinctive properties which could explain why certain activities – and only a few of them – have managed to obtain sufficient autonomy to organize and regulate their pursuits and their labor market. The original tension and classic debate between *profession* and *occupation*, or nonprofession, were revivified, put to the empirical test, discussed. By definition, the substantial results are variable, reflecting the diversity of the activities studied, located in contrasting positions in the scales of prestige, remuneration, expertise consecrated by training, collective organization, etc. On a theoretical level, this empirical work is diverse, but is most often part of the interactionist pole in the sociology of professions. Due to the very variety of the objects and occupational activities studied, they – most often implicitly but no less clearly – break with the functionalist perspective according to which only a limited number of activities reunite the characteristics legitimizing access to the status of *profession* (Carr-Saunders and Wilson 1933; Parsons 1939, pp. 34–49). The discriminatory use of the term profession appears largely inadequate to the empirical realities studied. Also, and often in an explicit manner, the frames of reference are inspired in interactionist perspectives which relativize or call into question its very principle: the distinction between *professions* and *occupations* (Hughes 1952, 1958; Becker 1962, pp. 24–46).

The variety of fields explored quickly made the profession's model (understood in the English language sense) and appear as too narrow a framework for organizing research, since this could not be restricted to the circle of canonical professions analyzed by the first sociologists of professions: mainly lawyers (Parsons 1952, pp. 370–386) and physicians (Merton 1957), which served as the basis for the functionalist modeling of the profession. We are aware of the low relevance of the distinction between *professions* and *occupations* in the French context, which is not only absent from ordinary language practices but also rarely used in sociological research (Menger 2003, pp. 1–26). The conception of professionalization as boundaries – door or bridge, barrier or pathway – between these two statutory categories is strongly dependent on an American or British institutional and political context, of little relevance for the countries in continental Europe. It nonetheless constitutes a powerful and solid matrix, since it leads to questions on the demarcation of this boundary, and therefore, the extent of professionalization. In other words, professionalization cannot be generalized because if it were a path open to every occupational activity regardless of its characteristic features, then the notion would tend to dissolve and blur any differentiation between professional entities, even when these are obviously not equivalent. On the contrary, if professionalization were reserved for a small number of métiers which fulfill functions deserving to be elevated to the rank of profession, then it would necessarily be limited to a few professional spheres constituting an elite or a privileged stratum in the productive system. These two polarized positions – the dynamic can be tightly restricted or potentially generalized – lead to a dilution of the fundamental signification of the concept, and of its analytical usefulness: a dynamic for producing differentiations and hierarchies, manufacturing distinctions and orders within the social division of labor.

The question of opening the access to professionalization – and therefore, to at least the theoretical possibility for any professional entity to follow a path toward professionalization – resonates with the empirical observations made over half a century ago by Wilensky (1964) when he noticed that a growing number of métiers and activities were trying to consolidate their statutory position, gain legal protection and control their work environment, in short, to professionalize themselves. In his functionalist conception, this meant standing out from the mass of occupations in order to acquire the elitist status of a profession. But the reasoning can be substantially softened, stating that an increasing number of professional groups seek to improve their position in the division of labor by accumulating the most diverse forms of recognition. For example, school heads sought to appropriate the managerial shift introduced vertically by the French National Education to gain room for maneuver in the management of their establishments, strengthen hierarchical power, negotiate an improvement in their remuneration and differentiate themselves from the teaching profession through more diversified recruitment (Pélage 2011, pp. 75–90; Barrère 2013). Likewise, the mediators responsible for managing family disputes prior to the intervention of lawyers emphasized the complexity of the relational skills required to conduct their activity and thus accumulate professional legitimacy and free themselves from established and more prestigious legal professions (Milburn 2002). Let us mention another professional sphere, that of nursing assistants: located in an intermediate position in the organization of paramedical work at the hospital, they take on the irksome tasks delegated to them by the nurses in order to enrich their own set of tasks, acquire a more honorable and rewarding role, and in certain cases, demand an increasing share of the work over the patients' bodies (Arborio 2012). Examples abound, such as that of magnetizers, traditionally subject to symbolic disqualifications because they base their practices on criticism of the superiority of scholarly knowledge over ancestral and analogical knowledge, but nonetheless manage to gain legitimacy by certifying their professional practices and paying taxes over the market relationships they are subject to (Charasse 2022).

The case of magnetizers leads to the limits of the field of sociology of professions, at least as it has been historically constituted, as they appear so distant from the ideas of métier, professional group and profession. However, like physicians, they carry out a consulting activity, relying on specific and esoteric skills (even if these are not integrated into a body of scientific knowledge). The service they offer has constitutive effectiveness and is the subject of a commercial exchange. In the same vein, we can take the example of prostitutes, a borderline case if there ever was one since trading their bodies, or selling paid sexual services, is generally categorized, independently of moral considerations, as an activity far removed from what we usually mean by work, and foreign to the world of professional groups. Prostitution-related activities are commonly stigmatized, accused of illegitimacy, subject to disapproval, considered deviant and sometimes illegal. But moral and legal orders do not hinder them from being considered an occupational activity, because apart from the fact of being remunerated, they involve a specific professional socialization, as all métiers do. Certainly, the absence of training premises does not translate a lack of abilities, because experience leads to the acquisition of ample know-how: adjusting the distance with customers, anticipating their reactions, identifying those who present a potential danger, earning respect and preserving safety, in short, mastering relational skills. Beyond that, these activity requirements are reflected in the discourses of prostitutes. Among those who vindicate prostitution as a métier (Pryen 1999), there is a distinction between the "professionals" who have acquired this know-how and those lacking said control, which are often included under other categories, such as "drug addicts". These positions, which reflect internal competition phenomena, are also attempts at regulation through the affirmation of a legitimate conception of the activity. At least in certain repressive circumstances, they can also lead to collective mobilizations for the activity to be recognized (Mathieu 2015) and considered equal to any other, as well as to have it regulated by labor law, a stepping stone toward professionalization.

This extension of the domain of professionalization corresponds to the starting point of Wilensky's argument, even if according to this author, many activities claiming statuary elevation do not meet the distinctive properties characteristic of professions: non-limiting exercise but pursued on a full-time basis, the existence of specialized training, the constitution of a body of rules organizing work, the founding of professional organizations, the formalization of a code of ethics and obtaining of an exercise monopoly. After all, this is his response to the central question he formulated himself: can everyone become a professional? By following his steps into the question but not in his way of solving it, we can remember that professionalization is an extensive social dynamic, which meets the aspirations of workers carrying out the most diversified activities and, in that sense, is potentially general. The field of education illustrates this wonderfully due to the variety of actors who contribute to educational work and seek to be recognized: the different traditional teaching métiers involve teachers, heads of establishment, inspectors, but also nonteaching staff who are increasingly diversified (Van Zanten 2001) and strive to find a place in the division of labor (Bois and Jacquot 2022), or even parents themselves, some of whom, particularly in the middle classes, claim their share of educational work (Ball and Vincent 2006). This generalization concerns professionalization as a process, not as an end result. It means that any group of workers identifying themselves with any activity or mastering a specific expertise are possibly engaged in a professionalization dynamic; in other words, this notion has a sufficiently positive connotation to function as a broad attractor. But in no way does this imply that the goal has been achieved and the aspirations satisfied, because the professionalization dynamic necessarily diffracts into unequal results and consequences.

The success of the theme of professionalization in the analysis of various activities does not lead to the identification of a global movement multiplying the number of métiers and other specialized activities, which would be increasingly valued one after the other. If researchers cherish the concept of professionalization, it is not only because they attach their questions to it, but mainly because it implies a form of valorization in itself: to turn into a professional has become an attractive goal in itself, even an obvious one, against which it is difficult to stand up... In the end, who would not wish to reach it, who would call into question such an undoubtedly desirable goal? We understand the interest of workers from all métiers in improving their status: autonomy in the definition and performance of their tasks, mastering the training conditions of the métier as well as the access rules to it, symbolic and material remuneration, recognition of consecrated expertise, etc. But this dynamic has heterogeneous translations, sometimes leading to explicit demands or negotiations with public authorities, sometimes remaining at the state of latent aspirations. On the other hand, professionalization does not lead to a leveling of activities and métiers but, as diffuse as it may be, it generates inequalities. This is why the question of the partition between the categories of professional entities – namely, between professions and nonprofessions – deserves to be further examined.

2.1.2. *Is there a professions perimeter?*

The emergence of the field of sociology of professions, in England in the 1930s and then in the United States, is inseparable from the identification of a specific object – professions – and a vision of modern societies in which these professions take on a central regulatory role by constituting an ethical and responsible alternative to the business world: "professions involve specialized intellectual technique, acquired through prolonged and formalized training, making it possible to provide effective service to the community" (Carr-Saunders and Wilson 1933, p. 284). From the very beginning this conception, which emphasizes professional ethics, a sense of service and scientific knowledge (Chapoulie 1973), leads to Malthusianism: all professional communities cannot be professions, because their eminent role is indexed to selective conditions. Delineating the perimeter of professions is therefore an original challenge. The introduction of the concept of professionalization translates this by showing that the list is not closed and the perimeter is not watertight.

In view of the distinction inherited from the sociology of English-speaking countries, *professions* are *occupations* which have managed to organize themselves to conquer exclusivity on the delivery of certain services, control the methods for acquiring expertise, define good working practices, gain autonomy in relation to the market and the state, resist the rationalizing forces encouraged by organizations and establish themselves as an economically efficient and morally desirable mode of regulation. They rely on their own professional organizations – orders, associations and unions – to build and consolidate their monopoly, their ethics and their ideal of service and to negotiate with public authorities for autonomy in the management of their professional territory, which they administer as a professional community (Goode 1957). This autonomy is a major resource for recognized professions (notably for the consulting professions which have long provided the main areas of sociological inquiry). They can assert the legitimacy of their expertise toward customers and users, and all the more so, when such expertise is associated with work on inference and practical wisdom (Champy 2009).

The late introduction of the sociology of professions in France after the 1990s quite logically highlighted debates on the perimeter of professions. The narratives in this field of research (Dubar and Tripier 1998; Champy 2009; Vézinat 2016) stage the fight between various schools of thought, in particular, the initial confrontation between functionalists and interactionists, and other alternatives. There is an obvious and precious educational value in this delimitation of the research field, revealing the importance of debates on professionalization understood as the differentiation and hierarchy of professional groups, in terms of recognition, autonomy and legitimacy. Still, the power of theoretical emulation of these descriptions is controversial, because the study of professional work in France has barely been structured around such debates. This is not to say that they were totally absent though, and the introductions to the history of the field have not been done in vain. The fact remains that the functionalist trend is little present in France, where it has been criticized (and even devalued), for its canonical conception of the profession, adhering to conditionality criteria, and struggling to give a clear status to professional rhetoric, which could end up being perceived as an endogenous attempt to legitimize successful professionalization, consecrated by a superior status. Conversely, and despite not being established as a paradigm of analysis, the interactionist trend has also found significant echo. Although some of its basic concepts (such as license, mandate, segmentation and negotiated order) are frequently used, this trend risks disqualifying the legal codification of professions de facto, while ignoring statutory inequalities not necessarily visible at the workplaces where it concentrates its analyses. We should also add that most of the time, the works quoted in support of a surpassing synthesis (such as those of Johnson, Larson, Abbott and Freidson) are classified under a third group which is so heterogeneous that its formation is not justified, other than to strengthen the supposedly structuring character of the principled opposition between functionalism and interactionism.

The interest in professions was sparked by a handful of works focusing on the dynamics of professionalization. They implicitly criticized the clear demarcation of a perimeter of professions and emphasized the plural and uncertain characteristics of the structuring of occupational activities (Tréanton 1960; Maurice 1972; Chapoulie 1973). But the debate was quickly closed, so much did it rely on categories foreign to the French sociology of the time. The analyses of social groups and occupational activities were organized around other categories, such as social classes, unions, métiers or socio-professional categories (Lallement 2007). And in the 1990s the development of the sociology of professions was not based on the delimitation of a specific object, such as a set of activities with such distinctive traits that they could be labeled as professions. Truth be told, the English language distinction between *professions* and *occupations* arouses empirical criticism to such an extent that the notion of profession has to be used cautiously. French sociologists often add a qualifier (established, liberal, organized, learned, accomplished profession, etc.), or use other terms, such as professional groups (Demazière and Gadéa 2009). The research field of the sociology of professions was not built on the basis of a differentiation – not even a hypothetical one – between certain activities which could be considered as professions, and others receiving a more common status. Opposing this differentiating perspective, it developed from fieldwork covering a large number of diverse activities. For example, in the contents summary of a first work aiming to structure this field of research in France (Lucas and Dubar 1994), we find a wide variety of "professional groups": physicians, customs officers, supermarket cashiers, engineers, nurses, teachers, drivers, train operators, consultants, farmers, trainers, etc. The comparison of these case studies stresses the nonlinear and non-univocal character of professionalization processes, which cannot, according to the coordinators, be reduced to an entry into the closed world of professions, but should rather be contemplated as a heterogeneous transformation in the recognition and valorization that professional groups benefit from.

This conceptualization marks a step aside from the logic demarcating the perimeter of professions. This gap is not only based on the broadening and diversification of fields, but also on the primacy given to research – and to the most varied fieldwork – fostering empirical situations which deviate from the canonical or ideal-typical definition of a profession. A few examples from the above-mentioned activities will help to illustrate this. Farmers claim to work for the common good, by feeding the population and maintaining rural areas. Although they have powerful professional organizations, they do not rely on the esoteric knowledge they monopolize and often appear as dependent large chemical firms (Joly 2001; Nicolas 2018). Nurses master specific expertise acquired in specialized schools, respecting a professional code of ethics. However, they are dominated by physicians who delegate the most irksome tasks to them, they are dependent on hospital bureaucratic organizations and have reduced autonomy in the exercise of their duties (Béghin-Do 2015; Douguet and Vilbrod 2018). Train drivers have an activity governed by numerous specific rules they do not enact because they result from regulations defined at higher levels. While their expertise guarantees traveler safety, they struggle to translate this goal into an ideal of service (Houriez 1994, pp. 275–288). Customs officers work for the community by positioning themselves as guarantors of fundamental collective rules and by fighting against fraud and crime. Nevertheless, their subordinate position in bureaucracies reflects the weight of the requirements to conduct their activity, even if part of it remains discretionary (Jacob 1995; Duprez 2000). The heterogeneity of the areas investigated, and the extent of deviations from the profession's ideal type in the situations studied keep us quite apart from the reasoning inherited from the tradition in English-speaking countries. The analysis of professionalization carves out its own path, far from the identification of salient, if not essential, features defining professions ("real" professions, we might even say). Such an analysis would be sterile or at least draining, since it would inevitably conclude with the shortcomings or incompleteness of professionalization, reducing the métier studied to the state of a semi-profession, a quasi-profession, a proto-profession, or why not a sub-profession, a quarter-profession, etc.

The conception of professionalization as the culmination of a differentiation process between two categories of entities – professions and nonprofessions – also comes up against both theoretical and practical difficulties, identified as early as the 1970s. The first relates to the formal definition of the boundary and the criteria used for delineating it. Even if Wilensky's proposals are the most widely disseminated, they are not the only ones. For example, four attributes of professions were also proposed around those times: specialized knowledge; a primary orientation to community interest rather than the profession's interest; a code of ethics; a sense of an accomplished work (a quality concern) (Barber 1963). But other characteristics were rejected because they were considered too widespread and insufficiently discriminating: corporate solidarity, specific socialization and lifestyle. We therefore understand that the identification of criteria enabling us to discern professionalization processes is a task with no end in sight (Jackson 1970) and one that has long been subject to debate (Greenwood 1957). The second difficulty lies in the relationship between the point of arrival and the process: how to measure and qualify the degree of accomplishment, knowing that it can be assessed in terms of distance traveled (and to be covered) and modulations depending on the chosen criteria. Here, the introduction of degrees of monopoly, external recognition, internal organization and knowledge esotericism (Turner and Hodge 1970, pp. 19–50) does not fundamentally change things, except that it turns boundaries into a vast intermediate area welcoming semi-professions (neither full professions nor occupations).

This invites the question on the need to draw a partition, regardless of the boundary's thickness, comprising the whole of occupational activities. Symmetrically, we can recognize the theoretical and empirical usefulness of not considering occupational activities in an indistinct manner. In order to describe, one has to classify, differentiate, organize. But criteriology (combining criteria to delimit a boundary) does not seem the right method to do so. The boundary is neither that of the ideal type (which measures differences between a standard and empirical cases), nor that of generalized equivalence, because the division of labor proceeds not only by differentiation, but also by hierarchization. The challenge lies in placing the culmination of professionalization at a distance, something which has been favored by the enrichment of the empirical bases and the affirmation of the concept of professional group.

2.1.3. *At a distance from the arrival point: professional groups*

The vitality of the field of study of professional groups in France quickly revealed the profession's model (in the English language sense) as too narrow a framework to grasp the process of division of labor, the differentiation of occupational activities and professionalization. Because according to this model, professions claim a monopoly of exercise, which they defend by controlling the content of the skills vindicated, the transmission of knowledge and the socialization of members, the ethical rules governing their implementation, and the social and economic value of their activity. A profession is a concept whose field of application is relatively narrow and targeted, where lawyers and physicians also occupy a prominent place (Dingwall and Lewis 1983; Dingwall 2008). And even keeping a distance from these emblematic figures, the concept applies to workers holding a high level of expertise, benefiting from great autonomy in the accomplishment of their tasks, forming a sort of professional elite at the top of the prestige or remuneration scale (Leichtet and Fennel 2001).

The introduction of the term "professional group" has made it possible to take into account various activities and métiers bottom-down of the social hierarchy, acknowledging varying degrees of organization or protection of the labor market (Chapoulie 1973). This paved the way for the renewal of theoretical questions. Instead of being considered as closed, protected and codified groups, professional groups were tackled as problematic entities: bringing together workers carrying out an activity with the same name, but whose social visibility, recognition and legitimacy were not assured, and whose internal cohesion, organization and strength were not established (Dubar and Tripier 1998). This perspective also resonates at an international level, where new expressions have emerged, such as *professional groups* or *occupational groups*, and more recently *professional occupations* (Evetts 2003, 2004).

Since then, the empirical expansion has been uninterrupted, to the point that it has become impossible to draw up a reasoned, if not exhaustive, inventory of the areas in which professional dynamics are studied. To fairly traditional fields, such as health (general practitioners, surgeons, nurses, psychotherapists, physiotherapists, but also pain physicians, acupuncturists or homeopaths), law (lawyers, magistrates, industrial tribunal judges, prosecutors, but also legal mediators, bailiffs, clerks), or the so-called liberal professions (accountants, notaries, bailiffs, auctioneers, insurance brokers), or education (secondary school teachers, primary school teachers, school heads, adult trainers, educational advisors, educational assistants for teachers), or even social work (social workers, child-minders, marital counselors, specialized educators, or even family aides, social mediators), are added diverse activities, such as the arts (theater actors, classical orchestra musicians, baroque musicians, jazzmen, circus artists or even art restorers), services to the individual (salespeople, delivery drivers, postal workers, teleoperators, bank advisors, postmen, funeral advisors or even online brokers), intellectual activities (journalists, urban planners, advertising creatives, industrial designers or even headhunters), small traditional métiers (fishermen, janitors, industrial workers, bakers or even dockers), public administration agents (police officers, professional advisors, pension fund liquidators, labor inspectors, tax agents, or even immigration clerks), atypical forms of activity (volunteer association leaders, locally appointed officials, high-level athletes, prostitutes, public writers and free software developers or even newspaper hawkers), etc.

The accumulation of empirical and monographic research focusing on diverse occupational activities has shown the way in which the mastery, control and work configuration challenges appear in an almost universal manner across all professional spheres and for all categories of workers. All of them face challenges in responsibility delimitation, knowledge recomposition, batch tasks modulation, collective strategies reorientation, in short, adjustments of their position in the division of labor. Therefore, does the challenge not come down to strengthening our position and accumulating resources until becoming a profession? A theoretical question thus resurfaces, together with the concept of professionalization: how to envision the processes, developments and changes theorized under the term "professionalization"? Do they adopt specific pathways for each occupational activity? Are they oriented toward a common direction, corresponding to professional stabilization or institutionalization? Can they be indexed to a degree measuring fulfillment or accomplishment? These questions point to a cumbersome paradox: the recourse to the notion of professionalization brings to the surface problems we used to associate with the notion of profession, and which we expected to solve just by adopting other categories of analysis. As a matter of fact, is the horizon of professionalization the formation of a professional group and the structuring of a profession? To what extent does the concept of professionalization provide a theoretical continuity enabling us to grasp degrees of advancement along a "normal" pathway, and in this case, what constitutes its real added value? On the contrary, to what extent do they correspond to a theoretical breakup, and in that case, what can be its significance other than the empowerment of a professional group and the recognition of a profession?

This points to a major difficulty in reasoning in terms of professionalization, to the risk of reducing the process to the outcome. The analysis in terms of processes has the advantage of breaking with static perspectives. However, implicitly or not, it risks modeling temporality in a historical sense, evolution in a given direction, progression graded in stages. Professionalization appears as a step forward, and sociologists risk having their activity reduced to the measurement of the progress covered, even if this means affirming the critical dimension of their point of view by showing that the actors concerned have over or under-evaluated progress along this path. Undoubtedly, the large-scale comparison and continuous empirical expansion have made it possible to harness this risk by emphasizing the heterogeneity of professional dynamics, resisting an interpretation in terms of fulfillment more or less conforming to a unique (professional) model, expressing multiple processes, a proliferation of logics not obeying any general norm. The notion of professionalization then serves to explore diversified, ambiguous and contradictory transformational movements professional activities undergo: emergence, identification, delimitation, categorization, legitimation, invalidation, erosion, segmentation, destruction and disappearance.

This extension of professionalization paths is interdependent on the conceptualization of the entities concerned by such processes: since these are potentially heterogeneous and fragmented, these entities are social forms which are themselves plural. This is how the very concept of professional group emerged for designating groups of workers who perform an identified activity bearing a specific name, and consequently, occupying a recognized place in the division of labor and benefiting from certain legitimacy, without introducing any hypothesis implying degrees of professionalization. Furthermore, these professional groups can be understood as fuzzy sets, subject to continuous changes in their activity and characterized by evolving contours. This perspective leads to placing the challenges of boundary delimitation at the center of professionalization analysis: not those which would separate professions from nonprofessions, but those identifying and delimiting the professional groups themselves.

2.2. Professionalization as friction at the boundaries

In a second analytical line, the culmination of professionalization processes does not appear at the center of attention, because it is not a question of separating categories of professional groups and observing if and how they reached the status of a profession. This statutory difference is placed in the background. It does not mean that it is being denied, but rather that the core of the analysis concerns the processes themselves. We could wonder about the interest of this position: why study a process while neglecting its point of arrival, its end result? The answer lies in the consequences of this theoretical choice – radical with regard to the tradition of professions analysis – on the conceptualization of professionalization processes. To consider that they have no culmination point amounts to breaking with the reasoning which conceives them as progression – or regression in certain cases – along a unidirectional path, based on indices and criteria measuring how these conditions are satisfied. This also comes down to conceiving professionalization as an ever unfinished and multidirectional process, expressed differently in a variety of professional spheres. It envisions it less as an internal dynamic of groups considered as entrepreneurs more or less effectively taking the reins of their own professionalization, than as a system of social relations whose alliances and conflicts delineate heterogeneous pathways and whose sequences are not based on a statutory progression vector.

While professionalization has been classically conceived as the production of boundaries between categories of professional groups, it is conceptualized here as work at the boundaries of groups, leading less to the accumulation of resources to improve their status than to the reconfiguration of groups, modifying their contours and characteristics. It is a more of a vector displacing boundaries between groups, rather than groups crossing a stabilized boundary. First, we will emphasize that all professional groups are surrounded, if not pinned down, by multiple actors claiming to influence their professionalization, and therefore causing tensions at their boundaries. Then, we will argue that even if wage labor (largely dominant) is the Achilles' heel of professionalization because it increases the permeability of professional groups to external influences, it does not necessarily signify deprofessionalization. Finally, drawing on the proposals of Andrew Abbott, who gives primacy to boundaries over professional entities, we will argue that every professional group constitutes professionalization.

2.2.1. *Tensions around professionalization*

Professionalization appears as a desirable goal, generally endowed with positive values, invested by workers who see it as a way of valorizing their activity. It also seems attractive to other stakeholders, such as recipients or regulators of the activity, who see it a lever for the production of professional and qualitative work, stable and predictable. But the picture is not so irenic, because while professionalization seemingly achieves consensus, it does so at the cost of a number of ambiguities and misunderstandings. Less than a circulating asset with the virtue of satisfying the convergent expectations of tradespeople, customers hiring them and the authorities organizing economic activities, professionalization can be considered as a floating signifier, onto which each category of actors invests in different meanings. This cannot be apprehended without taking into account the multiplicity of actors who gravitate around the boundaries of professional groups. The place of the latter – and even of *professions* – within societies is always caught in relationships of power, competition, struggle, negotiation, as stressed by the successive points of view adopted by Freidson (1986, 1994, 2001) on control by (the) professionals. For each stakeholder, professionalization is a space for projections and demands which, far from converging, often tend to collide. Because it expresses claims to define labor, and not only that, but also what professionals should or can be.

The strength and multiplicity of tensions affecting the definition of the missions of professional groups is a general phenomenon, more the rule than the exception. We can illustrate this through the example of three contrasting professional groups, not only by the positions they occupy in the division of labor but also by their degree of coalescence: physicians who have long been considered (Carr Saunders and Wilson 1933) as a typical case of an established profession and the subject of an impressive and probably unequaled amount of research (Bergeron and Castel 2018); teachers who, without being established as a recognized profession, benefit from a license and a mandate (Hughes 1958) as well as a large degree of autonomy in the practice of their métier, and have been studied in a regular and long-standing manner (Lieberman 1956; Dorros 1968; Geer 1968, pp. 221–234; Leggat 1970, pp. 153–177; Hoyle 1983, pp. 44–54); finally, building concierges, who are at the bottom of the professionalization scale in the sense explained in the first part of this text, occupy a poorly valued place in the division of labor and have aroused interest of a far more limited number of researchers (Bronner and Stébé 2001; Villanova and Bonnin 2006; Marchal 2006, 2007), with the exception of historical works.

What is a "good" doctor? The answers vary and depend on the positions occupied in relation to this activity: whether you are a doctor, a patient, a nurse, a hospital manager, the member of the patient's family, etc. The quality of a "good" doctor, of a "good" professional, gets caught up in a process of adjustment/misadjustment between competing – but not necessarily contradictory – claims to state what labor is, what it should be, and what its purpose could be. What is a "good" teacher? The teachers who perform this activity obviously have an answer to this question, as do the students, their parents, the school management, ministerial authorities, designers and publishers of school textbooks, pedagogues and didacticians, inspectors, teachers' unions, etc. Each point of view is indexed to a specific position in the world of education and associated with particular standards. What is a "good" building concierge? Certainly those who occupy these functions have an answer to this question, which they try to implement on a daily basis, but the direct recipients of their activity also have an answer, namely the residents and the owners or operators of the buildings, as well as other categories of workers whose activity has points of contact with that of caretakers, such as road workers, cleaners collecting household waste, gardeners maintaining green spaces around buildings, local traders, etc.

At the level of spaces and working premises, where occupational activities are effectively carried out, professionalization is omnipresent, at least in discourse and vindication/attribution actions. Not only do these acts contribute to configuring the daily lives of the workers concerned, but they also have an influence at a more global level on the definition of their mandate, the missions of the professional group and the regulation of their activity, in short, on professionalization. As a result, professionalization dynamics should not be interpreted in terms of increasing autonomy in the categories of workers, nor as a growing affirmation of expertise or professional groups. To be sure, they indicate that they are constantly confronted with other groups (of the most diverse nature), who also have an interest in conducting and defining what they do and who they are.

If we return to the case of physicians, it has been amply shown that they share the appreciation, as the definition, of their work with other actors: the implementation of health care protocols often requires the active cooperation of patients the success of medical work depends on. Patients also intervene in a more direct manner every time they require adjustments, make claims, or even file complaints. Caregivers also have an active role, when they invest in the relational dimension of medical work, often delegated for considering it to be dirty work. Hospital managers play a non-negligible role through resource allocation (or their reduction). Bureaucratic supervision can also exert increasing control through the automated feedback of medical procedures; public authorities can activate multiple regulatory levers, relating to the demographics of the professional group (through *numerus clausus* policies), remuneration (through pricing or even wages), assignments (including in the liberal profession model), etc. Even if they epitomize a recognized profession and as such benefit from large autonomy in the administration of their professional territory, physicians do not constitute a sovereign entity, but they must deal with other actors, are confronted with attacks or protests, are forced to negotiate compromises and must defend their position which would otherwise be eroded. In this sense, the relationship of physicians with a wide variety of entities without whom their mission could not be fulfilled and holding a position which is constantly at stake constitute a privileged point of observation to account for the trajectory of professional groups and to show the avatars of professionalization. The inventory of these relevant groups is always an unfinished task, so much does it depend on the events which sculpt the professionalization process. Without hesitation, we could add those actors who exert therapeutic practices not recognized as medical, not even under the label of alternative medicine (in France homeopathy, osteopathy, acupuncture and biorevitalization), and relegated to the margins or even to the field of magical practices (Chinese medicine, hypnosis, brain coaching, etc.), but who claim their expertise competes with recognized medicine, and therefore tend to threaten physicians by reducing their professional territory or by breaking into it to expand its perimeter, in both cases challenging its boundaries.

A similar reasoning can be applied to teachers. While this group is protected by a set of barriers to entry (such as expertise certified by training, or success in contests organized by public authorities), it is also the subject of interventions external to said professional group. This may just as well contribute to modifying their contours or influence the way in which they fulfill their missions. These missions are carried out in interaction with the students, and, through their (more or less cooperative or disruptive) attitudes, can influence the course of the activity and more broadly, the definition of teaching work, degrading it – if we follow the internal point of view – by imposing the need for discipline, for example. Other actors (such as inspectors and increasingly school heads who assess labor) also have a direct influence by providing recommendations to teachers, and promoting or disqualifying such or such teaching modality via the direct link between their evaluations and career development. We could also mention the parents of students who, while not forming a unified actor, share their opinions on a variety of scenarios, claiming to know how educators should behave (in the direct interactions with teachers, school boards where they are represented, in the public sphere through their federations). Teachers work under a wage regime – in or outside the civil service – and do not master the definition of their recruitment conditions, the groundings of the assessment of their activity, the progress of their careers, even if co-management with unions have long attenuated this phenomenon. Therefore, it is not only the autonomy which has long been recognized by them in their classes and in the sites where they work that can be contested. Beyond their license, it is also their mandate which can be contradicted and weakened, for example, by the development of *home-schooling* practices – whereby parents would replace teachers – or by the offer of complementary services to teaching which supposedly help students overcome limitations in the form of supplementary courses (even if it is partly the same teachers who provide them in a different setting, or when this task is entrusted to people who do not meet the required qualifications). Again, the boundaries of the professional group can be disputed, or discreetly displaced, to such an extent that professionalization can be found at the margins of the group in the relationships they establish with external actors on different scenarios.

Similar observations can be made about building concierges who, for their part, carry out an activity considered subordinate and who have not formed a collective organization capable of defending their interests and their conception of their work. This appears to be quite elastic and vulnerable to the intervention of external actors. In quantity as in quality, it receives the impact from residents who can cause damage and thereby deny or destroy it, or on the contrary, through their careful behavior, can contribute to the maintenance or cleanliness of the premises and acknowledge the concierge's role. The conditions under which they carry out their missions are also dependent on the resources allocated by the managers of trustees or real estate companies who hire concierges (as employees or service providers). The contours of their activity can also be the subject of negotiations with other professionals working within the same premises, initiating games of delegation of irksome tasks and the defense of prerogatives, particularly around cleanliness and waste management. Employers or public authorities also make attempts to redefine the métier, for example, by seeking to add social mediation missions, especially in so-called difficult neighborhoods, giving rise to more or less effective resistance strategies which oppose a purer conception of the métier. Building concierges are also engaged in continuous professionalization, enriched by a multiplicity of actors who intend to contribute to defining the métier's contours. The weak institutionalization of this professional group gives it few collective resources to influence these relationships, even if it is not completely helpless, since it can rely on collective agreements that provide the job descriptions or are inspired in the (imaginary?) solid foundations of a métier long rooted in history.

In this regard, each professional group finds itself in a unique situation: the range of actors who can modify the course of their activity and displace the boundaries of their professional territory is each time specific, as is the weight of the group in relation to these stakeholders. The inequalities are glaring here because the capacity for professionals to impose their own ways of defining their place in the division of labor and of fulfilling their role is tremendously variable. But the most prestigious, organized and recognized professions are not necessarily those who manage to escape the influence of the actors who gravitate around their boundaries. This is what the three examples described suggest.

Throughout the development of the sociology of professions, the medical profession is considered as an emblematic example – and without doubt, the major example – of successful professionalization. Barely had the functionalist trend pushed this line of analysis to the furthest when it came across the obvious limitations posed by the actual conditions of medical work, without nonetheless reducing it from the "institutional structure" it is inscribed in (Parsons 1955). Because these working conditions cannot be reduced to the singular consultation with each patient as in the private doctor model, where the practitioner is supposed to legitimately judge on the merits of the medical decision, from diagnosis to treatment (Freidson 1970; Pennec 2010, pp. 207–220). Even in that context, the therapeutic relationship cannot be assimilated to a relationship of domination, asymmetrical as it may be (Boltanski 1971). Above all, the practitioner's autonomy can be significantly limited, or contested by patients, and all the more so when cure is unattainable: in the case of chronic illnesses the doctor–patient relationship is part of a "negotiated order" (Strauss 1978), as patients can take an active part in treatment decisions and somewhat become specialists in their illness, or at least in the ways of handling it (Baszanger 1986). In addition, the singular consultation, which tends to be inscribed in a contractual model (Zussman 1997), is too narrow a framework to seize medical work when this is inserted in hospital bureaucracies, which act as a source of a second line of authority, that of the manager (Freidson 1984; Vassy 2000, pp. 71–86), or makes the physician's work inter-dependent with that of other professionals (Hall 1946; Herzlich 1973; Castel 2005). The conditions for implementing medical expertise are inseparable from a network of patients, parents, techniques and objects (Eyal 2013) to such an extent that it can be considered as a network attribute, not only of professionals.

In contrast, concierges occupy a marginal place in the studies on sociology of work. While they do not benefit from a professional status, they do appear to lack the capacity to organize themselves to claim it. Their professionalization, in the sense of proximity to the autonomous profession model, is minimal or even nonexistent. On the other hand, they experience multiple frictions at the boundaries of their professional territory. This, together with their place in the division of labor and the array of tasks assigned to them, are stabilized, but they are also subject to various movements propelled by their employers, their hierarchies and the recipients of their activity. However, this does not ipso facto deprive them from the ability to gain margins of autonomy for fulfilling their tasks, cooperate with residents or professionals working in contiguous territories to adjust their activity and reduce it or adjust it. The configurations may be conspicuously variable: some may favor their autonomy while others may constrain it, increasing or reducing the porosity of boundaries in their professional territory. As for teachers, they find themselves in an intermediate state of professionalization in the classic sense. While their profession does not benefit from the privileges equivalent to that of physicians, they do master specialized and scholarly knowledge, establishing a relational asymmetry with the recipients of their activity once their expertise has been recognized. This legitimizes their autonomy in the exercise of said activity, clearly identifiable in their working conditions, in the principle crowning them as the masters of their classroom, or even in the eminence of collegiality to regulate a wide set of aspects regarding their practice. But in no way do these protections prevent protests, both at the workplace and beyond, at the scale of a global society where educators are contested and criticized, and exposed to emerging forms of competition. As such, professionalization appears as an uninterrupted and uncertain movement.

2.2.2. *The risks of deprofessionalization?*

If professionalization is not a unidirectional movement, and a fortiori a cumulative one, how should we envisage the temporal forms it adopts: is it erratic, unpredictable and versatile; how can we account for the directions it takes, with its twists and detours, regressions and throwbacks? The argument of regression, or deprofessionalization, seems insufficient or simplistic, because it implicitly conveys a linear conception of professionalization, as an evolution featuring gains or losses, progress, immobility or setbacks. These movements can be considered in a more open way, as a set of displacements which are difficult to forecast, since they are the fruit of tensions at the boundaries of professional groups.

In multiple fields of activity, professional groups are part of a world of labor where their place, roles and attributes are put to the test by changes of all kinds, bringing weakening, strengthening and adaptation, in short, recompositions which do not only concern their autonomy for conducting their activity, but also engage their collective identity, cohesion and contours. For instance, the group of museum curators (Hénaut and Poulard 2018a, 2018b) is crossed by multiple dividing lines: their employment status (some being employees and others independent), the size and identity of the organizations where they work (national or territorial museums) and the missions assigned to museums (contemporary art, eco-museums, etc.). These cleavages also affect their missions and relationships with clienteles (the emphasis being placed on the exhibition's production, on broadening the audience, on museum activities, etc.), as well as the political and ideological orientations of the management and/or curators. When these dimensions are combined, they reveal segments within the same group, at the time they widen the gaps among those professionals performing the same activity, now bonded by "colleague-competitor" relationships. But, paradoxically, this segmentation contributes to increasing the group's vigilance against reform projects which could affect them, thereby consolidating the profession's boundaries in the face of adversity, as embodied by public authorities. Still, the pathway can be altogether different, as in the case of preventive archeology specialists. At the turn of the 21st century, they were subject to brutal changes in the regulation of their activity, as the opening to competition enabled private operators – as well as local authorities – to open up excavation sites and break the monopoly of the public operator (Bonnefon, 2024). This led to a diversification of employers and employment statuses, with a strong growth in temporary contracts and a deterioration of the wage ratio, weakening the group's cohesion despite the persistence of the same framework of activity and expertise. With the diversification of statuses, it is the working conditions and the quality of the work produced that were weakened. However, the development of contacts, particularly scientific ones, among preventive archaeologists of all statuses had a limiting effect on divisions and inaugurated a renewal of coalescence which prefigured a refreshed professionalization dynamic. In these two cases, we observe a composite and multidirectional professionalization, associated with group "re-stratification" (Kirkpatrick et al. 2009).

These two examples make it possible to identify a structural and structuring characteristic of a number of professional groups: the liberal practice, often related to professional status, occupies a marginal place, whereas wage labor prevails. On the contrary, this status is marginal in emblematic professional groups in France, notably medicine and law, even if it is established there: among lawyers the employment rate is around 5%, with strong disparities depending on specialties, with a strong dissemination in large international firms. Among physicians, progress in wages is difficult to assess accurately, but a growing number of communities are opening health centers in order to fight against medical deserts and offering to hire physicians there (who remain 20 times fewer in number than liberal practitioners), whereas 20% of young physicians prefer wage labor to freelancing (source: *Ordre national des médecins*; Conti 2019). Wage labor imposes additional constraints on the professional, the counterpart for the protection provided by the status. This in fact gives rise to an additional actor, namely the employer, who rightly formulates demands on the activity, reinforcing the population of those around professional groups, who are then confronted with managers, hierarchical agents and chief executive officers. In broader terms, wage labor can introduce rules which overflow the sphere of work/activity and influence the integration conditions of the professional group through recruitment practices, the professional territory through the hierarchy of missions and group cohesion through assessment schemes and career modulation (Boussard et al. 2010).

The acknowledgment of wage labor in the analysis of professionalization was progressive and, above all, was first interpreted as a threat to the autonomy of professionals and a degradation in the capacity of professional groups to administer their territory. Since the 1970s in the United States, this perspective has been oriented toward the analysis of relationships between professional groups and organizations, insisting on their conflicting nature. Indeed, the groups exposed to wage labor are heterogeneous. They differentiate themselves depending on their degree of structuring, the force of attraction of their organizations, the power of their representatives, cohesion in the defense of their interests, their capacities for collective struggle, or even their social prestige, the legal protections acquired, or the trust they inspire. But regardless of their empirical situation, wage labor and enrollment in more or less bureaucratic organizations could reverse the professionalization process, inaugurating its deterioration. Three major processes of this type have been identified. The thesis on *deprofessionalization* developed by Haug (1975) exposes how the rationalization of organizations, as well as the rise in consumer movements, led to subordinating professional autonomy to exogenous goals, depriving professionals from the power to set benchmarks for assessing their own work. This, in turn, weakened the professional group's capacity for action. It has also been argued that the trend inserting professionals into the wage system exposes them to an alignment of their condition with that of any worker forced to sell their labor force, thus supporting a form of *proletarianization* characterized by a rigid assignment of tasks by hierarchies and creating a "factory-like" scenario (Oppenheimer 1973, pp. 213–222). In the wage labor context, the relationship between professionalization and *bureaucratization* is described as inverted in most cases (Hall 1968). The inclusion of professionals in hierarchical lines would transform their work via a *bureaucratization* process (Scott 1966): their activity being planned by superiors who relay the purposes of the organization, control their results and disqualify their own rules and collegial norms.

Wage labor, which is the dominant status in post-industrial societies, very much like the French society which is described as a "wage society" (Castel 1995), is seemingly the Achilles' heel of professionalization, at least if compared with the liberal status, idealized and confined to a few rare activities. These analyses were rekindled with the progression of the *New Public Management*, which penetrated public organizations where many professionals toiled, and was expected to weaken the autonomy and power acquired by professional groups (Bézès and Demazière 2011). This reactivated an old polarity identified by research on the relationships between organizations and professional groups, opposing *management* and *professions*, or *managerialism* versus *professionalism* (Gleeson and Knights 2006). Other works called for overcoming essentialist or homogenizing visions of the professional reality (Evetts 2003, 2006, pp. 61–79) and for thinking more carefully about the relationships between organizations and professions (Ackroyd 1996; Kirkpatrick and Muzio 2011). On these aspects, the reader can refer to Chapter 4.

The wage labor status or enrollment in a bureaucratic organization does not necessarily threaten the professional expertise that characterizes work activities, but rather fuels collective consciousness, the sense of shared interests and collegiality. Because a number of professional groups integrated into organizations and working in the administration, for example, in the field of health care (notably the hospital sector), justice, police, education, higher education or even social action, appeal to specialized and codified knowledge, rely upon practical knowledge acquired through experience, know how to cope with the cases they are confronted with, refer to a shared conception of their function and recognize themselves in a collective identity defined by their specialty: their métier (Lipsky 1980; Sternberg and Horvarth 1999). However, dependence on their employer hinders the creation of professional organizations and the structuring of collective action oriented toward the group's endurance. This situation which distances them from the *professional status* model could be designated by new expressions such as *occupational professions* or *organizational professions* (Noordegraaf 2007). These expressions invite us to think about professional groups as a continuum of hybrid and mixed forms, rather than situate them within a metric of professionalization. This signals a rupture between professionalization and the professional status. Analyzing professionalization requires observing not only the internal dynamics of professional groups, but also their relationships with other actors (of which organizational actors are but one example, among political bodies, social movements, market operators, regulatory authorities, groups of all kinds), which flourish at their boundaries and have potential effects on them.

2.2.3. *Professional groups as professionalizations*

The configuration of professional groups is at the heart of Abbott's (2016) processual approach, as is his analysis of the "system of professions" (Abbott 1988). Fundamentally, this author proposed considering social entities (such as professional groups) as something secondary to boundaries and the delimitation processes establishing and displacing their contours, in short, to professionalization processes. This position was synthesized by a stunning formula: "we should not look for boundaries of things but for things of boundaries" (Abbott 1995a). This "*things of boundaries idea*" inaugurated a real theoretical counterpoint, rejecting the assumption that "the social world consists of fixed entities (the units of analysis) that have attributes (the variables) [which interact] to create outcomes, themselves measurable as attributes of the fixed entities" (Abbott 2001, p. 39). In contrast, entities, like professional groups, should not be considered as fixed realities, whose essence could be revealed by clarification analysis or sociohistorical purification, inferring certain attributes from them or passing them through the sieve of specific criteria. Instead of being at the origin of the processes which make them come about, they are the outcome, the product of a permanent work of adjustment and differentiation of boundaries.

While the theory is general, it is nonetheless relevant for the historical analysis of professionalization and the emergence of professional groups. The case study of social work in the United States at the end of the 19th century (Abbott 1995b) shows that the activities, services, tasks associated and coagulated by the delineation of a new boundary making social work appear as a visible professional space pre-dated the existence of organized groups or collectives claiming such place in the division of labor or such label. The gradual and slow crystallization of an area of specialized tasks took place without an institution or collective delimiting it, controlling it or labeling it. Those tasks were initially fulfilled by disparate categories of workers. They formed a vague, unstable and uncertain whole whose contours were gradually asserted and clarified, without such boundaries (yet) designating a precise entity or professional group. Those emerging boundaries produced uncertain, "differentiation sites" still under construction. As a result, a variety of workers were disseminated along both sides of such boundaries, differentiating them into categories which asserted themselves, whereas others faded, were recomposed or simply disappeared. The activities and workers located within the emerging perimeter resonated with one another, forming sets which appeared more and more significant, until becoming differentiated and designated by the (vaguely unifying or constitutive but still fragile at that stage) expression "social work".

Over time, a professional group came to exist through the articulation and consolidation of boundaries, which enabled the transformation of a "differentiation site" into a more visible, robust and enduring unit: in that case, social work, which no longer designated an uncertain phenomenon (as it had before), but a social entity having hatched along that process. Professionalization did not operate without the intervention of actors. But this is not a question of social work professions, since they were secondary to the connection of boundaries and did not yet exist as such. It was rather a question of charitable movements, government services and workers attached to varied tasks, which contributed to the coalescence of social work as a differentiated set of activities. However, this coalescence was partly contingent. While certain tasks were integrated into the area of activities still under construction, others were not. Such differentiation varied depending on country and time, to the point that, as Abbott has stressed, social work may not have existed as we know it today. It could also disappear as an entity, because entities are never stable and subject to perpetual change, to such extent that permanence is ultimately an "accidental outgrowth" (Abbott 2007, p. 8).

In line with his processual perspective, Abbott traced a concept of professionalization acknowledging social work, but without taking the form of an alignment with constitutive criteria. He showed that the regime of existence of social work evolved over the period studied, following an uncertain and specific path. First of all, it appeared as a poorly identified sphere of activity. Then, the very dynamic of the activity created a certain inertia, something to attach to, something which became more visible, something which showed signs of existence, something invested by various actors, and which became less and less uncertain. The existence of social work was thus affirmed, this "something" became a thing or an entity, endowed with increasing internal coherence and a certain endurance, until it became an institution with more clearly delineated and progressively protected boundaries, through formalized rules, enabling the development of specialized professional groups. His analysis of social work had the particularity of being historical, centered on the birth – and we could even say the gestation prior to the birth – of social work. But in other texts, he emphasized different moments of this continuous social process we have come to call professionalization. He illuminated the way in which professions are born, grow and die, split or regroup, adapt or innovate (Abbott 1988). We can therefore wonder – as Abbott does in the last paragraphs of his text – what happens to the theory of boundary work in the case where professions are constituted, when they are organized – de facto – entities or things. How can we maintain the idea of the primacy of boundaries over professional groups not when these are in their infancy, but when they have become solid entities, institutions? What happens with professionalization, then?

The scope of the theory does not reduce professionalization to a state where professional groups assert themselves and become objectified, manifesting the existence of an exclusive community whose members are united by collegiality relationships and whose access is regulated by mechanisms which precisely let boundaries be guarded. From the processual perspective, professionalization is provisional and unfinished. The analysis can be extended to all regimes of existence of professional groups, whatever the state of coalescence, visibility, recognition, weakening or decay which characterizes them. This means that professional groups (more or less constituted, coherent or contested) are constantly confronted with the challenge of their survival. Their future is established in relation to other entities, themselves more or less consolidated, and who somehow find an interest in professional services: competing, subordinate, complementary groups and audiences (clienteles, public authorities, certification agencies, the public opinion, etc.). These relationships can be extremely varied: conflict, predation, parasitism, threat, imitation, alliance, symbiosis, etc., and they centrally concern boundaries, which they potentially call into question, destabilize and displace to the point of weakening/consolidating differentiation sites whose metamorphoses mark the dynamics of professional groups, of professionalization.

Dependent on the systems of actors, professional groups cannot persist in an autarkic manner, withdrawing to the territory of their practice. These audiences are often decisive in the allocation of jurisdiction to the extent that they "serve as arbiters, because their own legitimacy has its roots outside the system of professions" (Abbott 2003, p. 30). Furthermore, and this opens up toward an unlimited diversification of the ingredients of professionalization, Abbott (2003, pp. 29–50) emphasized that those audiences are not homogeneous and uniform, but form ecologies linked to one another. Like professional groups, they are crossed by tensions and conflicts, operating by competing conceptions of their action in relation to such or such professional group, which increases the variability of relationships and interdependencies between professional groups and audiences, between professional ecologies and external ecologies. In this ecological and processual conception, professional groups, and more broadly, society, are never in balance; "conflict and unpredictability are the nature of social life" (Abbott 2006, p. 318), professional groups are dynamics or flows, in other words, professionalizations.

Regardless of the degree of thingness or endurance of professional groups – and *professions* have reached the highest degrees in this matter – boundaries are shifting, offering sites for observing professional dynamics. Professional groups are the product of history and social work along the boundaries having constituted them. As such, they cannot break free from those boundaries, they are inseparable from the audiences surrounding them, just as they are inseparable from work at their boundaries, and from their boundaries. Any professional group has the status of an unfinished process, a flow of events, partly controlled and partly unpredictable, and intertwining relationships at the boundaries, articulating variable combinations of potentialities, tensions, alliances, conflicts and threats. In other words, professional groups, whatever their degree of institutionalization, are professionalizations in themselves.

2.3. Conclusion

Finally, to seize professionalization in terms of the boundaries dynamics between professional groups invites us not to make assumptions on the solidity of such boundaries and on the consistency of the groups they contribute to differentiating. In that sense, professionalization is not the recognition of a professional group and its more or less opposed rapprochement with a statutory figure of reference – the *profession* – nor does it designate the capacity of a professional group to impose itself in the face of the demands from partners engaged in the normative definition of the activity: a job well done. It is a broader process. It designates the differentiation and continuous recomposition of occupational activities and the workers who perform them, while focusing on a division of labor that accentuates professional dynamics.

The concept of professionalization carries a strong ambition: to designate the division of labor, the continuous production of activities and groups as well as the recomposition of their relationships. Considering labor from this angle sheds light on fundamental social mechanisms affecting the whole labor market: (1) differentiation, through which social categorizations, the variable autonomy of groups and cross-dependencies including them in productive organizations and labor markets are recomposed; (2) coalition, through which collective identifications are formed, the defense of specific interests is organized, and multiple conceptions of activities are confronted or combined; (3) recognition, which supports self-realization, regulates work rewards and distributes differential values, creating more or less stable inequalities.

From this perspective, professional groups do not form a homogeneous whole – this is a mechanical consequence of the processual approach – but on the contrary, considerably differ by their degree of internal institutionalization, their capacity to dominate audiences, their mastery of selection at entry, their power to control their members, etc. In view of the perspective deployed here, no combination of characteristics constitutes a reference figure or a model of accomplished professionalization. Briefly stated, there is no accomplished professionalization, this being considered as intrinsically and irremediably provisional and unfinished. All professional groups, regardless of their visibility and identity, are confronted with the same challenge for asserting a place in an interrupted division of labor. Professionalization is not uniform, but on the contrary, it is distinctive and hierarchical, it produces multiple differentiations across occupational activities which concern the collective form of groups, the legitimacy of their expertise, the valorization of their work, the recognition of their usefulness, etc. Professionalization is not autonomous, but collective and disseminated, dependent on ecologies or systems of actors having an interest in the activity in question, and therefore working on the boundaries between this and the group. It is work – comprising struggles, alliances, construction of fortresses and offensives – at the boundaries of groups, which governs their reconfiguration and the modification of their contours, their legitimacy and their endurance. Professionalization concerns all professional groups, those which are in an embryonic state and relatively indistinct as well those which have become asserted, institutionalized entities. It follows heterogeneous paths in their outlines, their directions and their – always provisional – points of arrival.

2.4. References

Abbott, A. (1988). *The System of Professions: An Essay on the Division of Expert*. University of Chicago Press, Chicago.

Abbott, A. (1995a). Things of boundaries. *Social Research*, 62(4), 857–882.

Abbott, A. (1995b). Boundaries of social work or social work of boundaries? *Social Service Review*, 69(4), 545–562.

Abbott, A. (2001). *Time Matters: On Method and Theory*. University of Chicago Press, Chicago.

Abbott, A. (2003). Écologies liées. In *Les professions et leurs sociologies*, Menger, P.-M. (ed.). Éditions de la MSH, Paris.

Abbott, A. (2006). The concept of order in processual sociology. *Cahiers Parisiens*, 2, 315–345.

Abbott, A. (2007). Mechanisms and relations. *Sociologica*, 2, 1–22.

Abbott, A. (2016). *Processual Sociology*. University of Chicago Press, Chicago.

Ackroyd, S. (1996). Organizations contra organizations: Professions and organizational change in the United Kingdom. *Organization Studies*, 17(4), 599–621.

Arborio, A.-M. (2012). *Un personnel invisible. Les aides-soignantes à l'hôpital*. La Découverte, Paris.

Bajard, F., Crunel, B., Frau, C., Nicolas, F., Parent, F. (eds) (2018). *Professionnalisation(s) et État. Une sociologie politique des groupes professionnels*. Presses Universitaires du Septentrion, Villeneuve d'Ascq.

Ball, S. and Vincent, C. (2006). *Childcare, Choice and Class Practices: Middle-Class Parents and Their Children*. Routledge, London.

Barber, B. (1964). Some problems in the sociology of professions. *Daedalus*, 92(4), 669–688.

Barrère, A. (2013). Un management bien tempéré : l'expérience des chefs d'établissement de l'enseignement secondaire français. *Éducation et sociétés*, 32, 21–34.

Baszanger, I. (1986). Les maladies chroniques et leur ordre négocié. *Revue française de sociologie*, 27(1), 3–27.

Becker, H.S. (1962). The nature of a profession. In *Education for the Professions*. National Society for the Study of Education, Chicago.

Béghin-Do, M. (2015). Tuteurs et changement. La professionnalisation des infirmières en question. Continuité, transformation ou mutation d'une profession. PhD Thesis, Université de Toulouse, Toulouse.

Bergeron, H. and Castel, P. (2018). *Sociologie politique de la santé*. PUF, Paris.

Bézès, P. and Demazière, D. (2011). Introduction : new public management et professions dans l'État : au-delà des oppositions, quelles recompositions ? *Sociologie du travail*, 53(3), 293–305.

Bois, G. and Jacquot, L. (2022). L'autre travail éducatif. Personnels non enseignants et division du travail. *La nouvelle revue du travail*, 20, 6–19.

Boltanski, L. (1971). Les usages sociaux du corps. *Annales ESC*, 26(1), 205–233.

Bonnefon, H. (2024). L'esprit du patrimoine et les archéologues à l'épreuve de la libéralisation des fouilles préventives : un groupe professionnel dans les réformes. PhD Thesis, Sciences Po, Paris.

Bonnet, M. (2015). Le métier de l'aide à domicile : travail invisible et professionnalisation. In *Perspectives en clinique du travail*, Clot Y. (ed.). Érès, Toulouse.

Boussard, V., Demazière, D., Milburn, P. (2010). *L'injonction au professionnalisme*. Presses Universitaires de Rennes, Rennes.

Bronner, G. and Stébé, J.-M. (2001). Les gardiens-concierges : d'un métier à une profession. *Espaces et sociétés*, 105, 211–228.

Carr-Saunders, A.M. and Wilson, P.A. (1933). *The Professions*. Clarendon Press, Oxford.

Castel, R. (1995). *Les métamorphoses de la question sociale. Une chronique du salariat.* Fayard, Paris.

Castel, P. (2005). Le médecin, son patient et ses pairs. Une nouvelle approche de la relation thérapeutique. *Revue française de sociologie*, 46(3), 443–469.

Champy, F. (2009). *La sociologie des professions*. PUF, Paris.

Chapoulie, J.-M. (1973). Sur l'analyse sociologique des groupes professionnels. *Revue française de sociologie*, XIV(1), 86–114.

Charasse, F. (2022). Magies de la modernité. Comment le magnétisme en France et le chamanisme au Pérou sont devenus des pratiques légitimes. PhD Thesis, EHESS, Paris.

Combessie, P. (2009). *Sociologie de la prison*. La Découverte, Paris.

Conti, T. (2019). Pourquoi les jeunes généralistes de la Somme choisissent-ils l'exercice du salariat exclusif ? Thesis, Université de Picardie Jules Verne, Amiens.

Danet, F. (2012). *La médecine d'urgence. Vers de nouvelles formes de travail médical*. Érès, Toulouse.

Demazière, D. and Gadéa, C. (2009). *Sociologie des groupes professionnels. Acquis récents et nouveaux défis*. La Découverte, Paris.

Demazière, D. and Pélisse, J. (2022). Entre organisations et professions. In *Une société d'organisations*, Borraz, O. and Musselin, C. (eds). Presses de Sciences Po, Paris.

Demazière, D., Roquet, P., Wittorski, R. (eds) (2012). *La professionnalisation mise en objets*. L'Harmattan, Paris.

Dingwall, R. (2008). *Essays on Professions*. Routledge, London.

Dingwall, R. and Lewis, P.S.C. (1983). *The Sociology of the Professions: Lawyers, Doctors and Others*. Macmillan, St Martin's Press, New York.

Dondeyne, C. (2002). Professionnaliser le client : le travail du marché dans une entreprise de restauration collective. *Sociologie du travail*, 44(1), 21–36.

Dorros, S. (1968). *Teaching as a Profession*. Ohio University Press, Athens.

Douguet, F. and Vilbrod, A. (2018). *Les infirmières libérales ? Une profession face aux soins à domicile*. Seli Arslan, Paris.

Dubar, C. and Tripier, P. (1998). *Sociologie des professions*. Armand Colin, Paris.

Duprez, D. (2000). Modes d'apprentissage et formes de savoirs. Une comparaison police-douanes en France. *Déviance et société*, 24(1), 3–25.

Evetts, J. (2003). The sociology of professional groups: New questions and different explanations. *Knowledge, Work and Society*, 1, 33–55.

Evetts, J. (2004). The sociological analysis of professionalism: Occupational change in the modern world. *International Sociology*, 18(2), 395–415.

Evetts, J. (2006). Organizational and occupational professionalism: The legacies of Weber and Durkheim for knowledge society. In *Sociological Essays for a Global Society*, Marcuello, C. and Fandos, J.L. (eds). Prensas Universitarias de Zaragoza, Saragosse.

Eyal, G. (2013). For a sociology of expertise: The social origins of the autism epidemic. *American Journal of Sociology*, 118(4), 863–907.

Freidson, E. (1970). *Profession of Medicine*. Harper & Row, New York.

Freidson, E. (1986). *Professional Powers, a Study of Institutionalization of Formal Knowledge*. University of Chicago Press, Chicago.

Freidson, E. (1994). *Professionalism Reborn. Theory, Prophecy and Policy*. University of Chicago Press, Chicago.

Freidson, E. (2001). *Professionalism. The Third Logic*. Polity Press, Cambridge.

Geer, B. (1968). Occupational commitment and the teaching profession. In *Institutions and the Person*, Becker, H.S., Geer, B., Riesamn, D., Weiss, R.S. (eds). Aldine Publishing, Chicago.

Gleeson, D. and Knights, D. (2006). Challenging dualism: Public professionalism in 'troubled' times. *Sociology*, 40(2), 277–295.

Goode, W.J. (1957). Community within community: The professions. *American Sociological Review*, 22(2), 194–200.

Gottmann, J. (1980). Les frontières et les marches. Cloisonnement et dynamique du monde. In *Geography and Its Boundaries. In Memory of Hans Boesch*, Kishimoto H. (ed.). Kummerly+Frey, Bern.

Greenwood, E. (1957). Attributes to a profession. *Social Work*, 2(3), 35–45.

Guibert, P. and Lazuech, G. (2010). Parcours professionnels et professionnalisation des enseignants du secondaire : des régularités du social aux trajectoires singulières. *Recherches en éducation*, 8. doi: 10.4000/ree.4469.

Hall, O. (1946). The informal organization of the medical profession. *The Canadian Journal of Economics and Political Science*, 12(1), 30–44.

Hall, R.H. (1968). Professionalization and bureaucratization. *American Sociological Review*, 33(1), 92–104.

Haug, M.R. (1975). The deprofessionalization of everyone? *Sociological Focus*, 8(3), 197–213.

Hénaut, L. and Poulard, F. (2018a). Faire groupe au sein d'un groupe : la structuration des segments professionnels. Le cas des conservateurs de musée en France. *SociologieS*. doi: 10.4000/sociologies.8798.

Hénaut, L. and Poulard, F. (2018b). Réformes publiques et autonomie des groupes professionnels : le jeu des segments. *Revue française de sociologie*, 59, 259–284.

Herzlich, C. (1973). Types de clientèle et fonctionnement de l'institution hospitalière. *Revue française de sociologie*, 14, 41–59.

Houriez, C. (1994). Les agents de conduite de la SNCF : formation et socialisation. In *Genèse et dynamique des groupes professionnels*, Lucas, Y. and Dubar, C. (eds). Presses Universitaires de Lille, Villeneuve d'Ascq.

Hoyle, E. (1983). The professionalization of teachers: A paradox. In *Is Teaching a Profession?*, Gordon P. (ed.). University of London Institute of Education, London.

Hughes, E.C. (1952). The sociological study of work: An editorial foreword. *The American Journal of Sociology*, LVII(5), 423–427.

Hughes, E.C. (1958). *Men and Their Work*. The Free Press, Glencoe.

Jackson, J.A. (ed.) (1970). *Professions and Professionaliszation*. Cambridge University Press, Cambridge.

Jacob, F. (1995). Les douaniers français aux frontières de la sécurité intérieure. *Déviance et société*, 19(4), 339–354.

Joly, N. (2001). La professionnalisation des 'nouveaux paysans' : identités et rapports au territoire. *4ème congrès international, Actualité de la recherche en éducation et formation*, Lille.

Kirkpatrick, I. and Muzio, D. (2011). Professions and organizations: A conceptual framework. *Current Sociology*, 59(4), 389–405.

Kirkpatrick, I., Kragh Jespersen, P., Dent, M., Neogy, I. (2009). Medecine and management in a comparative perspective: The cases of England and Denmark. *Sociology of Health and Illness*, 31(5), 642–658.

Lallement, M. (2007). *Le travail. Une sociologie contemporaine*. Gallimard, Paris.

Lang, V. (1999). *La professionnalisation des enseignants : sens et enjeux d'une politique institutionnelle*. PUF, Paris.

Le Bianic, T. and Vion, A. (2008). *Action publique et légitimité professionnelle*. LGDJ, Paris.

Leggat, S.G. (1970). Teaching as a profession. In *Professions and Professionaliszation*, Jackson, J.A. (ed.). Cambridge University Press, Cambridge.

Leicht, K.T. and Fennell, M.L. (2001). *Professional Work. A Sociological Approach.* Blackwell, Oxford.

Lieberman, M. (1956). *Education as a Profession.* Prentice Hall, New York.

Lipsky, M. (1980). *Street-Level Bureaucracy. Dilemmas of the Individual in Public Services.* Sage, New York.

Lizé, W. and Naudier, D. (2015). Intermédiaires, professionnalisation et hétéronomisation des champs artistiques. In *Bourdieu et le travail*, Quijoux, M. (ed.). Presses Universitaires de Rennes, Rennes.

Lucas, Y. and Dubar, C. (eds) (1994). *Genèse et dynamique des groupes professionnels.* Presses Universitaires de Lille, Villeneuve d'Ascq.

Mallochet, G. (2007). A l'école de la détention : la socialisation professionnelle des surveillants de prison, trajectoires et expériences dans l'institution carcérale. PhD Thesis, CNAM, Paris.

Marchal, H. (2006). *Le petit monde des gardiens-concierges. Un métier au cœur de la vie HLM.* L'Harmattan, Paris.

Marchal, H. (2007). Les gardiens d'immeubles : le présent conjugué au passé. *Formation emploi*, 97, 95–107.

Mathieu, L. (2015). *Sociologie de la prostitution.* La Découverte, Paris.

Maurice, M. (1972). Propos sur la sociologie des professions. *Sociologie du travail*, 14(2), 213–225.

Menger, P.-M. (2003). Introduction. In *Les professions et leurs sociologies. Modèles théoriques, catégorisations, évolutions.* Éditions de la MSH, Paris.

Merton, R.K. (ed.) (1957). *The Student Physician. Introductory Studies in the Sociology of Medical Education.* Harvard University Press, Cambridge.

Michon, S. and Ollion, E. (2018). Retour sur la professionnalisation politique. Revue de littérature critique et perspectives. *Sociologie du travail*, 60(1). doi: 10.4000/sdt.1706.

Milburn, P. (2002). La compétence relationnelle : maîtrise de l'interaction et légitimité professionnelle. *Revue française de sociologie*, 43(1), 47–72.

Nicolas, F. (2018). Entrer en agriculture biologique : sociologie politique d'une professionnalisation sous contrainte (1945–2015). PhD Thesis, Université de Bourgogne France Comté, France.

Noordegraaf, M. (2007). From 'pure' to 'hybrid' professionalism. Present-day professionalism in ambiguous public domains. *Administration and Society*, 39(6), 761–785.

Oppenheimer, M. (1973). Proletarianization of the professional. In *Professionalization and Social Change*, Halmos, P. (ed.). Keele University Press, Keele.

Parsons, T. (1939). The professions and the social structure. In *Essays in Sociological Theory*. The Free Press, New York.

Parsons, T. (1952). A sociological looks at the legal professions. In *Essays in Sociological Theory*. The Free Press, New York.

Parsons, T. (1955). *Éléments pour une sociologie de l'action*. Plon, Paris [Original published 1951].

Pélage, A. (2011). Chefs d'établissement : la fabrique des cadres. In *Manager ou servir*, Lamarche, T. and Bechtold-Rognon, E. (eds). Éditions Syllepse, Paris.

Pennec, S. (2010). La médecine générale, une 'profession consultante' entre profession savante et prestations de services. In *Singuliers généralistes. Sociologie de la médecine générale*, Bloy, G. and Schweyer, F.-X. (eds). Presses de l'EHESP, Rennes.

Perkin, H. (1988). *The Rise of Professional Society*. Routledge, London.

Pichonnaz, D. (2017). *Devenir policiers*. Antipodes, Lausanne.

Pruvost, G. (2007). *Profession : policier. Sexe : féminin*. Éditions de la Maison des Sciences de l'Homme, Paris.

Pryen, S. (1999). *Stigmate et métier. Une approche sociologique de la prostitution de rue*. Presses Universitaires de Rennes, Rennes.

Salaün, M. (2008). *Le métier de facteur à l'épreuve des nouvelles organisations du travail à La Poste*. PhD Thesis, Université Paris-Est.

Scott, R. (1966). Reactions to supervision in a heteronomous professional organization. *Administrative Science Quarterly*, 10, 65–81.

Simmel, G. (1988). Le pont et la porte. In *La tragédie de la culture*, Simmel, G. (ed.). Payot, Paris [Original published 1909].

Sternberg, R.J. and Horvarth, J.A. (eds) (1999). *Tactic Knowledge in Professional Practice*. Lawrence Erlbaum, Mahwah.

Strauss, A. (1978). *Negotiations: Varieties, Contexts, Processes and Social Order*. Jossey-Bass, London.

Tréanton, J.-R. (1960). Le concept de 'carrière'. *Revue française de sociologie*, 1(1), 73–80.

Turner, C. and Hodge, M.N. (1970). Occupations and professions. In *Professions and Professionaliszation*, Jackson, J.A. (ed.). Cambridge University Press, Cambridge.

Van Zanten, A. (2001). *L'école de la périphérie. Scolarité et ségrégation en banlieue*. PUF, Paris.

Vassy, C. (2000). Carrières et relations de pouvoir chez les médecins hospitaliers : une comparaison européenne. In *Professions et institutions de santé face à l'organisation du travail*, Cresson, G. and Schweyer, F.-X. (eds). Éditions de l'ENSP, Rennes.

Vézinat, N. (2016). *Sociologie des groupes professionnels*. Armand Colin, Paris.

Villanova, R. and de Bonnin, P. (2006). *Loges, concierges et gardiens. Enquête en Europe : Paris, Londres, Barcelone, Milan, Oslo*. Créaphis Éditions, Saint Etienne.

Wilensky, H. (1964). The professionalization of everyone? *American Journal of Sociology*, 2, 137–158.

Zussman, R. (1997). Sociological perspectives on medical ethics and decision-making. *Annual Review of Sociology*, 23, 171–189.

PART 2

Professionalization, the Rise of Professionalisms

3

New Requirements and Standards at Work: Driving Forces and Limitations to the Recomposition of Professional Autonomy and Responsibility

Arnaud MIAS
IRISSO, Paris-Dauphine University, France

Work organizations have undergone major developments in recent decades. The transformations embraced, however, varied in their forms and rhetoric, share the same credo: overcoming a disqualified Taylorism. New managerial and organizational models have been proposed, which have inspired not only initiatives or experiments, but also forms of mimicry. Far from a scenario replacing one rationalization formula for another, we are witnessing strong inertia, hybridizations and the maintenance (and even the exacerbation) of a diversity of organizational forms. Although there is an affirmation of new productive paradigms, their coherence is lacking and they leave ample leeway to various managerial and professional strategies (Segrestin 2004). These transformations are accompanied by new expectations, requirements and standards at work, in short, by new more or less shared rules (or proposed by ones for the others) about work and the conditions for its exercise (quality, professional practice benchmarks, etc.). For Wittorski (2009, pp. 781–793), these transformations point to a professionalization dynamic that manifests the development of a culture of results, power decentralization and "an (apparent) calling into question of external regulations, delegating onto individuals

Encyclopedia of Professionalization,
coordinated by Didier DEMAZIÈRE and Richard WITTORSKI.

the responsibility for greater efficiency of the activity, together with a stronger individualization in the relationship to said activity".

This professionalization movement appears all the more powerful as few work situations seem to escape it at present. Far from being limited to a few highly qualified professional groups, the demand for professionalism seems to have become a universal standard of contemporary capitalism. Such an extension borders on indeterminacy: the very notion of professionalization loses its clarity and accuracy, as it designates an ever wider range of organizational phenomena. This is because managerial discourse tends to take up the characteristics and values having enabled professions to differentiate themselves from other workers (see the first part of this encyclopedia): chiefly, an ethic of responsibility and autonomy in the organization of labor.

This evidence an "injunction to professionalism", which is becoming widespread. Boussard et al. (2010) have defined this injunction to professionalism as "a set of imperatives externally imposed over workers, subject to new definitions of their missions and confronted with efficiency standards governing their activity" (p. 13). Contemporary organizations, both public and private, have become the source and support of a call for professionalism, whose effects are ambivalent. While it acts as a promoter of greater autonomy and responsibility at work, it seemingly opposes the traditional professionalization logic and the conquest of autonomy by professional groups who have acquired control over their work and the regulation of its exercise (division of labor, definition of missions and tasks, recruitment conditions, career rules, training content, etc.). This loss of power on the part of *professionals* is the threat the contemporary organizational transformations seem to carry. In any case, this is the hypothesis defended by Freidson (2001) in *Professionalism, The Third Logic*: today, customers, on the one hand, and managers, on the other hand, seem to be responsible for monitoring the work of these professionals. These assaults on *professions* (sometimes described as "neoliberal" attacks) are expressed as increasing competition and a requirement to evaluate the activity, an injunction to *be held accountable* in relation to customers or managers. A whole set of management schemes seek to assess the work of professionals through various indicators. By setting targets or priorities for them, they weaken the ability of professionals to freely deliberate on these ends and the means to achieve them. The rise of heteronomous control comes at the expense of professional self-regulation.

This thesis of a decline in professional autonomy must be tempered, because a more complex representation of labor scenarios has emerged from empirical research. Hybridization in the modes of regulation is frequently observed, rather than pure substitution, as highlighted in the work edited by Kuhlmann and Saks (2008) on the "governance" of health professions in different national contexts. Professional groups can still fundamentally influence the nature and form of public control and new governance practices. In addition, multiple professional segmentations make the effects of these managerial transformations highly variable, and therefore, complex to grasp: vast research inspired in the ecological approach developed by Andrew Abbott insists "less on the risks for *professions* as on the dynamics at the heart of the perpetual recomposition of the division of labor, of the activity's modes of exercise and their legitimacy" (Dubar et al. 2015, p. 283).

The inclusion of numerous professional groups in a wage labor relationship and the fact that their activity increasingly takes place in large organizations invites us to take into account the dependence of these workers on their employer. As customers or employers, these organizations can dominate certain fields of activity, like consulting, auditing or legal services firms which have especially drawn the attention of researchers (see Kipping et al. 2006, pp. 153–165; Faulconbridge and Muzio 2008; among others). New paths to professionalization are emerging, which make these large companies the privileged place and support for professionalization. This involves strategies and organizational schemes. At the time, it requires the initiative and involvement from these companies. It is in this sense that Daniel Muzio et al. (2011) spoke of "*corporate professionalization*". What we observe in these expert professional groups can be said for many other professional groups. Several works have brought to our attention the deployment of an injunction to demonstrate professionalism, which tends to affect all sectors of activity and categories of workers. From a Foucauldian perspective, Fournier (1999) identified a new "professionalism rhetoric", which she interpreted as a vector for the self-discipline of professional workers and for remote control. This call for professionalism could be based on a disciplinary logic: the standards this discourse promotes function as a disciplinary tool that enables managers to exercise control over expert work through "responsibilization" (Grey 1997). This call for professionalism is based on a manipulation of identities, a form of "subjection" in the sense of Michel Foucault, implying both subordination and security, dependence and power. For her part, Evetts (2003) has suggested speaking of *corporate professionalism*. She interprets contemporary transformations less in terms of "deprofessionalization" than as the source of renewed tensions between different forms of professionalism. This organizational professionalism takes shape through a whole set of competing schemes and values, which can replace professional values and regulations (Evetts 2011). These schemes and values are proposed, even imposed, externally to the professional group (*from above*): training and certifications escape the initiative and control of the professional group; managerial and/or bureaucratic control replaces peer control; performance targets are set by others, as are budgetary restrictions and the injunctions to account for the activity and its results. In this sense, professionalism does not designate the reasoned reliance on a body of knowledge, know-how and values a profession has acquired, but the capacity of workers to incorporate requirements external to their own activity, such as compliance with announced deadlines and costs.

In this chapter, we propose to interpret the development of these new standards at work as a tension between different ways of conceiving the occupational activity, its standards and values. This tension straightforwardly affects the core of professionalism because it is likely to recompose the way in which the quality of the work accomplished is defined, together with the regulatory mechanisms to ensure that a job is done well. The destiny of professional autonomy is then at stake, that is to say, the concrete weight of professional groups in the definition of their own labor "understood both as a set of practical activities to be performed and the resources to be summoned accordingly, and as a set of pursued goals and associated significations" (Boussard et al. 2010, p. 159). Such an approach to professionalism invites careful consideration of the evolution of autonomy, not only in fulfilling our work, but also in defining its requirements. This autonomy has a collective basis and is not reserved for established professions. If we follow Florent Champy's proposals, the question would be less oriented to professional autonomy, than to the social conditions conducive to "practical wisdom", understood, in the Aristotelian sense of the term, as the possibility of deliberating to "find relevant solutions to singular and complex problems for which science and routines are insufficient" (2012, p. 205). In other words, are new organizational standards likely to affect the way in which workers prioritize several purposes inherent in their activity? This type of questioning invites us to closely watch the effects of transformations in organizational contexts on the work activity itself (beyond the recomposition of power relations) and to evaluate professional autonomy in relation to the material organization of work, the division of labor and the criteria for evaluating professional work.

In this chapter, we will present the results of research regarding the effects of the development of these new requirements and standards at work in various situations and working contexts. After putting forward the way in which statistical surveys shed light on these general developments (section 3.1), the analysis will focus on four normative transformations acting as a clear example of contemporary managerial and organizational developments. First, we will address the notion of project management, often considered as the pivot of new organizations in the working environment (section 3.2). We will then analyze the tensions which arise from the efforts to standardize work, paying special attention to the implications of lean management (section 3.3). We will then look into public organizations and the effects of New Public Management (section 3.4). The last part of this chapter will question the scope and challenges of experimenting with new forms of work governance and discuss this concept (section 3.5).

3.1. Greater autonomy in the face of increasing work requirements?

National and European statistical surveys[1] provide fairly accurate information on the way in which organizational transformations have led to changes in work requirements and autonomy over the past 30 years. A general movement of work intensification has been observed since the end of the 1990s (Gollac and Volkoff 1997). In concrete terms, this intensification has been characterized by an increasingly constrained work pace. This increase in time pressure has different sources, which indicate both a plurality of requirements weighing on work and their accumulation (see Table 3.1). We notice the simultaneous progression of an immediate dependence on colleagues, production standards having to be met within the day or the hour, external requirements (most of the time demanding an immediate response), constraints or permanent surveillance exercised by the hierarchy as well as technical constraints (like the automatic shift of a product or piece, or the self-rolling of a machine), to mention just a few. After a pause in their intensification, these constraints were reinforced at the turn of the 2010s. During that period there was also a slight increase in the share of employees declaring that their work rhythm was imposed by computerized control or monitoring.

Gollac and Volkoff (1997) showed that this intensification of work resulted from an accumulation of constraints against the backdrop of a progressive boundary erasure between the industrial world and the world of services. While commercial constraints were imported into the industrial world, service-related work became increasingly exposed to rhythm constraints that had long reserved for industrial work in the past. A rationalization movement, with various components, affected all services (Gadrey 1994), sometimes even taking the form of automation (Bernard 2012). For Askenazy (2004), it is the organizational and managerial innovations developed in the 1980s–1990s which combined their – sometimes contradictory

1 Working Conditions Surveys have been conducted by Dares (the French Ministry of Labor) since 1978 with a representative sample of all employed workers, in order to produce a concrete description of work, its organization and its conditions from various angles: working rhythms, physical efforts or risks incurred, arduousness, cooperation, conflicts, etc. The European Working Conditions Surveys (EWCS) have been carried out every five years since 1990 by the European Foundation for the Improvement of Living and Working Conditions. The latest survey carried out in spring 2021 was based on a sample of over 70,000 employed persons, spread across 36 European countries.

– effects to maintain this dynamic of work intensification and accumulation of rhythm constraints. From this various research emerged the idea of a general dissemination of organizational models, participating in the erasure of boundaries which long structured the division of the world of labor, not only between industry and services, but also between the private sector and public services. The resulting work intensification movement affected all sectors and socio-professional categories. A recent statistical analysis has identified this movement in the hospital professions over the last 20 years (Benallah and Domin 2017).

Work pace imposed by...	1984	1991	1998	2005	2013	2016
Immediate dependence from colleagues	11	23	27	27	30	29
Production standards to be satisfied within the day	19	38	43	42	46	48
Including those to be satisfied within the hour	5	16	23	25	27	29
An external requirement	39	57	65	65	69	70
Including those demanding an immediate response	28	46	54	53	58	58
Permanent surveillance exerted by the hierarchy	17	23	29	30	31	31
The automatic displacement of a product or a component	3	4	6	25	35	35
The self-rolling of a machine	4	6	7	7	8	8
Other technical constraints	7	11	16	7	7	7
Computerized control or follow-up				16	17	17

Source: Dares, DGAFP, Drees, Insee, Working Conditions Surveys. According to Bèque and Mauroux (2017).

Table 3.1. *Various requirements that weigh down the pace of work*

For some time, the intensification of work was accompanied by greater autonomy at work. In a certain light, we can interpret this increase as a responsibilization maneuver and a source of additional pressure in itself: the fact of having to solve problems and contingencies on their own and of having to take action while reconciling contradictory demands provokes tensions at work. But at the same time, this progression of autonomy can be read as a form of compensation for the increase in work requirements. The job strain analysis model developed by Robert Karasek, which today serves as the basis for numerous diagnoses of psychosocial risks, articulates three dimensions: the “psychological demand”, “decision latitude” and “social support”. According to this model, problematic situations for people’s health stem from an imbalance between these dimensions.

According to the results of the Sumer survey (*Surveillance médicale des risques professionnels*) conducted by the Dares (French Ministry of Labor), one of whose modules includes the Karasek questionnaire, the general situation of executives (a high level of psychological demand combined with a high degree of decision-making latitude) is in this regard less worrying than that of employees in the administration and services (Amira and Ast 2014).

Since the turn of the 2000s, statistical surveys have revealed a continuous decline in autonomy and leeway, for all socio-professional categories (see Table 3.2). Fewer employees today "have no deadlines or can modify their deadlines", or can "choose for themselves how to achieve the targets fixed" than they did 20 years ago. However, this development is far from uniform because at the same time we perceive an increase in the proportion of administrative employees and unskilled workers who do not always apply instructions, or who do not receive them altogether. Besides, more and more employees (especially among employees in commerce and services), state that their work enables them to learn new things.

Tables 3.1 and 3.2 highlight contrasting developments in work requirements and autonomy at work. But as Table 3.2 shows, the differences in working conditions between socio-professional categories have remained relatively unchanged over the last three decades. When autonomy declines, it is approximately in the same proportions for all the socio-professional categories (PCS, French nomenclature), with a few exceptions.

The European Working Conditions Survey (EWCS) raises the question as to a possible French singularity in relation to these developments in autonomy and work requirements. We first observe that French employees occupy an average position in relation to the different dimensions of work intensity (Inan 2014). They often declare that their work pace depends on customer requirements, the direct control of their bosses or production or performance targets. They also state that they frequently have to interrupt a task to carry out another unplanned one. On the contrary, they are slightly less concerned with accomplishing complex tasks, meeting specific quality standards, or self-assessing the quality of their work. Dynamically, in recent years, work intensity has remained stable in Europe, with an increase in English-speaking countries, the Baltic countries and southern countries and a slight decrease in continental European countries (Eurofound 2021). This means that France is following a trajectory of work intensification closer to that of the United Kingdom compared to Germany or Belgium.

		Executives	Intermediate professions	Administrative employees	Commerce and services employees	Qualified workers	Nonqualified workers	All
Choosing the best way to reach our goals on our own	1991	91.2	89.1	80.3	77.7	77.1	59.4	**79.9**
	1998	96.3	92.0	84.8	82.0	82.1	66.1	**85.1**
	2005	92.9	87.2	77.5	77.0	78.6	59.5	**80.8**
	2013	90.1	83.1	78.3	76.9	75.6	63.0	**79.8**
	2016	89.2	81.9	74.5	73.4	72.5	58.4	**77.3**
Solving issues on our own, at least in certain precise cases	1991	89.2	79.2	64.1	54.9	58.9	39.4	**64.8**
	1998	89.6	81.9	70.6	66.1	65.7	48.9	**71.7**
	2005	86.3	78.9	68.7	65.6	66.8	47.9	**71.1**
	2013	82.4	75.3	65.7	61.8	63.6	52.5	**69.1**
	2016	81.3	73.8	61.2	63.4	62.0	50.1	**68.1**
Not having to meet deadlines or being able to change existing ones	1991	74.8	73.1	63.9	63.3	52.4	49.4	**62.9**
	1998	76.3	73.5	66.7	69.7	56.1	54.4	**66.8**
	2005	75.1	72.3	65.9	73.9	58.4	55.7	**68.4**
	2013	69.2	65.3	59.1	69.2	53.1	53.2	**63.4**
	2016	67.8	67.4	53.9	64.6	52.6	50.1	**61.9**
Not being obliged to apply directions or not receiving them	1991	78.9	70.3	54.8	51.7	52.0	38.7	**58.0**
	1998	81.3	70.0	57.3	56.3	58.8	45.5	**62.4**
	2005	81.2	70.6	61.7	61.3	58.6	44.9	**64.8**
	2013	81.4	69.5	63.8	60.2	60.4	53.9	**66.5**
	2016	81.3	71.4	64.4	58.2	56.0	51.6	**65.6**
Learning new skills	1991							
	1998							
	2005	93.5	87.8	79.1	62.8	65.9	54.3	**75.6**
	2013	94.7	86.2	77.8	65.3	67.8	53.8	**76.9**
	2016	93.9	89.1	83.3	71.9	68.9	53.8	**79.7**

Source: Dares, DGAFP, Drees, Insee, Working Conditions Surveys. According to Bèque and Mauroux (2017).

Table 3.2. *Evolution of autonomy and leeway*

The degree of autonomy at work appears to be significantly lower in France than in the rest of Europe for certain socio-professional categories and sectors of activity. Declarations on the possibility of "influencing decisions which are important for one's work" is particularly relevant: it concerns 51% of French employees, 61% of employees in the European Union, and even 84% in Scandinavian countries. French employees are also a little less "able to choose or change their work pace or speed" (61% compared to 66%) or "to change their working methods" (60% compared to 63%). They are also rarely "consulted before the targets of their work are set". Finally, only 68% say that their job "implies learning new things", compared to 90% in Scandinavian countries. Most importantly, the dynamic analysis highlights a French development contrary to the general trend in Europe, revealing an increase in autonomy between 2005 and 2015 (Eurofound 2021).

In addition to comparing responses between countries, data from the EWCS make it possible to analyze the way in which the different variables are combined within the framework of distinct organizational forms. Based on data from the 2000 European survey, Edward Lorenz and Antoine Valeyre (2005) carried out an ascending hierarchical classification of European employees based on 15 variables. This analysis led them to distinguish between four forms of work organization. Two forms are relatively traditional and little affected by recent innovations: "simple structure" organizations (labor is not highly constrained and not very repetitive, but relatively monotonous and controlled by direct supervision of the hierarchical superior) and Taylorian organizations, characterized by repetitive and monotonous tasks, significant work pace constraints and low procedural autonomy. E. Lorenz and A. Valeyre also draw our attention to two forms which emerged from this period of rapid transformation in work organizations. Lean production organizations combine teamwork, versatility, quality management and tight flows. They expose employees to unforeseen learning and problem-solving situations, but provide relatively little autonomy and generate multiple pace constraints. Finally, "learning" organizations share several common features with lean organizations (such as teamwork, self-monitoring of the quality of work, learning situations and contingencies resolution), but stand out for their strong autonomy at work and relatively few constraints on pace. The new work requirements take shape in various productive configurations outlining at least two regimes of labor standards in contemporary organizations, which radically combine autonomy and work intensity. In this European group, French organizations show a relatively marked appetite for lean production than for learning instances, as well as by the relatively stronger maintenance of Taylorian and simple forms of work organization.

3.2. Project, network and trust: new labor standards?

According to statistical data, a general movement of work intensification was, for some time, accompanied by a greater recourse to worker autonomy. These statistical findings were amply corroborated by qualitative surveys covering varied working contexts and professional groups. They invite us to consider a profound and extensive transformation of work organizations, which some interpret as the emergence of a new "model", with novel values and work standards. Autonomy at work seems to be the subject of contradictory, or at least ambivalent, developments, which sociologists have attempted to translate through the use of oxymorons: Coutrot (1998) and Appay (2005) spoke of "controlled autonomy", while Lallement (2010) coined the term "autonomy under constraint".

Among the many proposed interpretations of these transformations, the work of Boltanski and Chiapello (1999) paid particular attention to the shift in work standards, as evidenced by the analysis of management literature. The authors focused on the ideology which justifies employee commitment within an economic system characterized by capitalist accumulation. This ideology articulates individual considerations (the reasons for becoming engaged in a capitalist enterprise) with general considerations (how this commitment serves the common good). Boltanski and Chiapello showed how a new ideology asserted itself in the post-1968 period to counter "social" and "artistic" criticism, promoting risk, mobility and activity, and advocating for the liberation of the individual, who could thus express creativity and fully flourish. The figure of the "manager" was brought to the fore, as the one who has not only the vision, but also numerous relationships and can shift from project to project thanks to their mobility and networking skills. The manager is also the one who leads through charisma and motivates people to work with them on the basis of trust (passage from control to self-control). The manager is the one who must lead those who work along with them in the same direction, while encouraging the emergence of their creativity. A new justificatory apparatus is set, called "the projective city" (*cité par projets*), with its "common higher principle", to judge acts, persons and activities. It designates the ability to generate projects or to become integrated into existing ones, conceiving life itself as a series of projects. In this view, it is important to develop activities, voluntarily engage in projects and acquire the capacity to insert ourselves into networks to explore them in order to place ourselves in a position that will generate still new projects. The abilities to connect, communicate, coordinate, adjust to others, trust and generate confidence are positively praised. As we can see, through their analysis of management literature, Boltanski and Chiapello make project management the archetype of the new forms of work organization, which call on the initiative, creativity and employee commitment.

Echoing this perspective, Courpasson (2000) has called into question the liberal appearance of contemporary organizations, which present themselves as spaces of decentralized power, cooperation, participation and responsibility. By analyzing the case of bank modernization, he denounced the muffled violence underlying the appeal to commitment, which actually functioned as a scheme for bringing individuals into compliance. Far from being able to make creative use of their skills, employees would be compelled to show their compliance with managers' requests. Project teams would be based on "constrained solidarity" and principles of surveillance and struggle for positions, which grant social conformity. In that sense, project management would be no more than the new clothes of domination at work.

The interpretations we have just presented share the trait of making project management the pivot of new work organizations. While project management appears to be a powerful vector of new standards and work requirements, its real scope still has to be measured. Let us now explore part of the empirical work which shows what project management does to work and professional autonomy in specific working contexts.

3.2.1. *Professionals threatened by project management*

At the beginning of the 1990s, a first literature review (Ford and Randolph 1992) identified the advantages and disadvantages of matrix organizations[2]. On the one hand, such a model opens new communication channels (horizontal or lateral), which increase the frequency of exchanges and the volume of information the organization can process. On the other hand, it is a source of conflict not only within the organization (between functional managers and project managers), but also among the individuals summoned to cooperate despite their having different perspectives on their work, the temporalities it involves or the organizational targets it tackles. In a broader sense, it can also be costly both for the organization in terms of meetings, delayed decisions and data processing, and for individuals, due to role ambiguity, conflict and stress.

As part of an ethnographic investigation, Flamant (2002) analyzed the effects of the matrix organization on labor relations in the aerospace industry. He showed that this type of organization causes permanent confrontation between different production logics, responsibility being trusted to several categories of actors. Working relationships are marked by the more or less harmonious meeting of

2 A matrix organization designates a series of cross-functional organizational overlaps which create multiple chains of authority, gathering employees in teams to work on specific tasks for limited periods of time.

different types of responsibilities (e.g. development on one side, production on the other), without any formal hierarchical relationship between them, but intertwined by the need to cooperate, as a result of their interdependence. In such a matrix organization, each employee is dependent on two sources of authority: the hierarchical superior (who ensures the technical supervision of a team built around a "job", technical expertise, a field of specialty) and the project, program or product manager(s), who set targets, define requirements and allocate resources. Everyone is placed in a work context that makes their activity impossible without discussion between the different authorities they depend on. In theory, this form of organization tends to considerably reduce autonomy, insofar as each person's work is dependent on the intervention of a third party, with cooperation being compulsory. It manifests through the establishment of decentralized coordination and decision-making bodies, supposed to take charge of solving inevitable tensions and dysfunctions, concomitantly holding everyone accountable. Opposite the Taylorian model's perspective, the matrix organization invites employees to agree to develop in an uncertain environment and to establish negotiation as the rule. In practice, however, Nicolas Flamant has shown that it is the heads of programs who dominate working relationships, insofar as they are representatives of the customer's requirements. The project's logic tends to prevail over the métier's logic. And technical managers express the strongest criticism against matrix organization because it places them in a position as simple performers. Added to this is a feeling of dispersion due to the multiplication of actors and meetings and the complexity of decision-making processes. What emerges is the impression of being more of an executor of processes decided elsewhere than an actor directly involved in production. Employees would be deprived of their capacity for deliberation and decision-making. Further research has shown that, far from the sometimes expected close-knit conviviality, the experience of the project team is primarily that of discomfort with permanent negotiation (Midler 1993). Furthermore, such an organization entails the obligation for project collaborators to "trace" their work and make it permanently available to their colleagues.

This tension between the métier and project-based organization, constitutive of matrix organizations, was also observed by Coutant (2018) a few years later, but in renewed terms, in the aeronautics industry. In a pendulum swing, after the turn of the 2010s, the company's management tried to revalorize the "profession" to compensate for the excesses of matrix logic. But this managerial promotion of the métier was inherently ambivalent. The question was not to reopen margins for professional autonomy, but, by making a rationalizing gesture, to standardize and regulate the design activity more firmly, subject as it was to numerous uncertainties, as expert work generally is. For R&D engineers, this was translated as more constraints at work, rather greater autonomy. The engineers mentioned feeling confined to artificial compartments, preventing them from evolving in what they designated as their "métier", in a sense quite distinct from that of management. The managerial valorization of electronics and software professions was short-lived because it contradicted the professional conception of such professions and could be interpreted by the designers it was addressed to as an initiative which relegated them to the bottom of the symbolic and power hierarchy: "software is considered not as a profession in which one makes a career but as an entry-level profession in a company ensuring relative labor security and career prospects for young engineers who then seek work with systems manufacturers" (Coutant 2018, par. 55). Designers strategically fled toward more valued métiers, such as systems engineers, or toward managerial rather than technical positions. In doing so, the managerial construction of the profession maintained the dynamics at work in industrial engineering which pushed engineers out of the most technical activities by promoting generalist skills.

If project management struggled to deliver on its promises to engineers, what can be said about the professionalization of project managers themselves? From a critical and Foucauldian perspective, Hodgson (2002) analyzed the development of project management as a discipline in the double sense of the term: on the one hand, the construction of a body of knowledge and expertise underlying this managerial field, and, on the other hand, the deployment of disciplinary schemes for training and control. The body of knowledge tends to embody and reproduce a form of technicist and instrumental rationality. Managerial techniques have to improve the calculability and visibility of expert work, making direct control possible. Project management is thus presented as a set of universal and neutral techniques, suitable for all types of activities and sectors. Its development largely depends on the representation of the knowledge system it relies upon as an objective, coherent and effective discipline. Based on a survey in two British financial institutions, Hodgson showed that project management has significant effects within these organizations, imposing a form of discipline both on the recipients of such schemes and on their promoters and managers. By means of different disciplinary technologies, a specific form of rationality is instilled in employees' subjectivities. Project management thus appears as a powerful means for intensifying the control exercised over employees, not only thanks to the levels of visibility and calculability it enables, but also through the inculcation of a reflexive form of self-discipline. According to Hodgson, this reveals the dual nature of project management, combining bureaucratic logic and behavior autonomization. However, in a subsequent publication, Hodgson (2005) strongly nuanced the subjective influence of project management. He described project managers as maintaining an ambivalence, even a distance, toward their expected professional role. On frequent occasions, the values and standards of professionalism and project management are mocked and parodied in the daily practices of employees and managers alike. Hodgson interpreted those jokes as a reflection of the difficulties for matching professional doctrine with concrete achievements, having to cope with the complexity and unpredictability of project work. The procedures prescribed by project management are far from being respected, without this fundamentally raising any problems or conscience issues:

> steps are omitted, documentation is overlooked, reports are skipped and best practice ignored in the everyday conduct of project managers and team members, without this appearing to call into question respondents' sense of their own professionalism (Hodgson 2005, p. 63).

3.2.2. *The promises of the Agile method put to the test by employee commitment*

Digital technologies have provided powerful support for operationalizing project management. Digital service companies (formerly SS2I) have formed a particularly favorable receptacle to this managerial innovation, endowing it with an undoubtedly less bureaucratic, more flexible form which leaves ample room for individual initiative and autonomy. At the turn of the 2000s, recognized IT developers promoted the "Agile method", built as a reaction to a form of project management deemed too bureaucratic[3]. This method is based on a critique of the control exercised by information systems management through the imposition of standard sequences of actions, which allegedly promote anticipation and favor coordination. Its promoters have stressed the numerous unforeseen events and contextual elements which make any planning impossible. The Agile method, on the contrary, emphasizes the values of collegiality, mutual adjustment and rational democracy, based on the idea that better results can be obtained by promoting flexibility, commitment within tight and independent and self-organized teams and face-to-face dialogue. It is today instrumented through technical schemes, among which the most used is the "Scrum" methodology, which allows for regular internal communication through short iterations, during which the team autonomously takes charge of the resolution of production problems in an adaptive and recurring manner. A prototype is quickly deployed, and then corrections, additions and new features are regularly made, based on user practice and requests. One of the strengths of this method lies in its ability to incorporate suggestions made within the multidisciplinary team (designers, developers and testers) to maintain flexibility while placing more responsibilities on the shoulders of employees, entrusted with the power to

---

3 The "Manifesto for Agile Software Development" (2001) was based upon four values: "We are uncovering better ways of developing software by doing it and helping others do it. Through this work we have come to value: individuals and interactions over processes and tools; working software over comprehensive documentation; customer collaboration over contract negotiation; responding to change over following a plan".

redevelop projects. They have the feeling of operating within a democratic framework and with significant leeway. From their point of view, they are not just implementing instructions but can actually summon their knowledge to propose and develop new ideas. However, as Hodgson and Briand (2013) showed, the values promoted by the Agile method come into tension with the managerial practices, which tend to limit their effectiveness. Based on a survey carried out in a video game development studio in Canada, they noticed the persistence of a hierarchical conception of decision-making power, the "*team leader's*" viewpoint tending to prevail over others, to the point that decisions largely mirror power hierarchies. Furthermore, project teams are strongly constrained by specifications imposed on them by the company's management, for example, in terms of quality standards, resource allocation or the designation of certain members, including the heads of project. In other words, while developers enjoy a great deal of autonomy, they do so within a framework where decision-making and monitoring mechanisms are defined and implemented by authorities outside the team. The power transference to the project team is tightly limited. Finally, certain team members, such as artists and animation specialists, who have a more individual conception of creation, do not play the project management game: they do not communicate the information needed for team coordination (they do not update the "*backlog*", that is to say, the list of functionalities still requiring work to fine-tune the targeted product). This undermines the scheme's capacity to permanently inform everyone about the progress on each task. In turn, this leads managers to return to a more direct form of control over the activity of those employees.

Boboc and Metzger (2020) studied the implementation of the Scrum method in the IT division of a large information and communication technology company. They too noticed the reluctance of some actors to play the transparency game required by the Agile method. For them, "versatility under the constraint of cooperation" is not self-evident. The fact of making someone's activity, their intermediate productions, but also their tinkering and tweaking visible, presupposes a particularly high degree of mutual trust: they only engage in it expecting others will do the same, and confiding that they will not exploit what is shared in that way for personal purposes. However, the context of individualized career management tends to undermine the groundings of this trust. Boboc and Metzger also showed that the implementation of the Scrum method has little to do with the rigorous application of a managerial recipe. On the contrary, it is a source of tensions with existing organizational modes. It first gives rise to adaptation to the company's structures, insofar as part of the tasks are subject to automation, at the time interoperability and security constraints are imposed. It then makes place for non-negligible tinkering, so as to cope with the paradoxical need to rely on supervision for solving the difficulties of the method's local application, while advocating the absence of hierarchy. The research conducted by Boboc and Metzger also lets us question the effects of the importation of the Agile method on the autonomy of IT professionals in a very large company. A mixed picture emerges. On the one hand, IT professionals find the resources to engage in their activity: the fact of working in a small team where they gradually become more versatile can provide a feeling of regaining control over the product of their activity and give meaning to their work. On the other hand, organizational and hierarchical constraints strongly limit their leeway, making the implementation of the Agile method very costly, particularly in terms of coordination. For the authors, these constraints can ultimately compromise the destiny of an organizational innovation that extensively relies on employee commitment: "this tension between *relative autonomy* within small teams and *illusory autonomy* on a larger scale, can lead to frustration and discouragement, thereby reducing initiative" (ibid., paragraph 70).

3.3. Professionalization or standardization? New forms of work standardization

Project management as in the Agile method seems to paradigmatically embody the new work standards, together with their contradictions and paradoxes, through a call to be autonomous and demonstrate professionalism in a job. Other managerial innovations present their ambition for control and standardization more openly. This is the case of quality management, which experienced rapid expansion in Western economies after the 1990s. We will see that *lean management* also took this path, albeit in a more ambivalent way.

3.3.1. *Quality management, between standardization and individual responsibilization*

Starting from the principle that it is necessary to identify failures and drifts ("non-conformities") before they materialize as "non-quality" – which is the case when goods or services do not meet standards, expectations and/or specifications – total quality management intends to engage all employees in the continuous monitoring of the entire productive process. This is translated by the elaboration of quality systems, which explain in detail the working procedures and resources to be used for addressing every event (scheduled or not) likely to affect the productive activity. In this view, quality management manifests as greater codification of the activity and more transparency in the organization. But at the same time, it promotes the autonomy of employees, by encouraging them to identify and to solve the problems which arise during their activity, on their own. It purports a self-control logic associated with a collective control of collective activity, far from management based on constraint and the decomposition of activity into workstations, or broken down into separate tasks (Segrestin 2004). Even if the procedural injunctions are reinforced, these can only produce effects through initiatives close to the activity. This evidence a consolidation of control, even an excess of bureaucratic zeal, involving an increased recourse to autonomy.

By engaging localized and tacit self-organization skills, the company's management seeks to promote both self-discipline and the appropriation of production and performance standards. The targeted organizational model deviates from more bureaucratic approaches, insofar as it intends to act more directly on employee subjectivity. By disseminating the problem-solving function, managers engage this subjectivity with the intention of making it an effective instrument of self-discipline at the service of the organization's goals. In other words, the appeal to professionalism (somehow involving mental manipulation) could be made for tightening up control. Some case studies, among the first published (see in particular Wilkinson and Willmott 1995), have clearly portrayed quality as a discrete way of intensifying work and strengthening managerial control, rather than autonomy. Far from encouraging cooperation and collective participation, quality management reinforces individual responsibility. Other empirical research avenues have nonetheless highlighted the incomplete fulfillment of such a managerial program. By studying the implementation of quality management at a large retail bank, Knights and McCabe (2000) detected the maintenance of a great deal of autonomy and an ability to resist such control. While quality management certainly transforms the subjectivity of employees, this does not mean that it eradicates all forms of resistance. The authors particularly emphasized the numerous situations leading to unsolvable dilemmas and the confrontation with contradictory demands. The ability of employees to resist managerial control resides precisely in these moments when control mechanisms are inconsistent or contradictory. For their part, McArdle et al. (1995, pp. 156–172) revealed that the employees studied were perfectly aware that quality had intensified their work, but they also perceived it as something enriching, not necessarily as a process that should be opposed. In the field of chemistry, Campinos-Dubernet and Marquette (1999) showed that the ISO 9000 certification process could support the formalization and valorization of tacit know-how into explicit skills: this certification explicitly recognizes that operators possess a form of expertise over their activity. Paradoxically, the writing of work procedures, which tends to homogenize the way of working, was not perceived by operators as something to endure. They rather interpreted it as an "ordering" of practices making it possible, among other things, to clarify the responsibilities of each worker while offering a representation of the entire production process, helping to perceive the impact of their own work on the activity of others. The writing of procedures would therefore bring out hybrid rules which, despite being written and imposed, would directly stem from operator practice. Observing the deployment of quality standards in an industrial waste processing company, Cochoy et al. (1998) offered a more nuanced interpretation. Admittedly, the codification effort modifies working relationships: through the collective writing process, it associates employees more directly with "organizational work" generally reserved for management staff. But standardization places everyone in tension between the access to formal recognition of their field of responsibility and exposure to reinforced control based on this recording of work practices. In the workshops of the automobile industry where this quality management is deployed, Rot (1998) observed a rearrangement of rules and the development of informal workshop conventions, far from any passive submission to control. The contradictory implementation of the principles of traceability (reporting all problems encountered) and responsibility (taking charge of quality control for ourselves) causes points of tension, which are solved by circumventing traceability schemes and replacing them by the local and informal construction of quality by the operators themselves.

Beyond the strict industrial framework where it emerged, the requirement to respect quality standards gradually extended to professional spheres, where they standardized work activities. Standards for good medical practice were developed, which modified the grounds for legitimacy in medical practices and affected the relationships between physicians, providing them with new tools to judge and classify themselves (Castel and Dalgalarrondo 2005). In nonclinical testing laboratories for chemicals, Lamy (2016) observed the use of "Laboratory Best Practices" (LBP) and the effect of this standardization on innovation inherited from scientific practices. According to the author, while the burden of these LBP does affect the smooth running of scientific practices, it does not always or necessarily lead to a weakening of innovation capacities. Improving and adapting the standards is possible, provided that a discussion takes place between quality managers and scientists.

Stevens (2010, pp. 41–50) emphasized the effects of market standardization on the work of computer scientists in R&D for a computer manufacturer. In the 1960s and 1970s, those professionals benefited from relative professional autonomy, enabling them to organize their activity fairly freely, to negotiate their working conditions and to keep their distance from hierarchical injunctions. After the 1980s (with the rupture represented by the invention of microcomputers), there was a weakening of the professional criteria for defining the activity ("making something beautiful") in favor of organization above all ("doing something useful"), which was soon replaced by commercial logic ("making a standard"). Useful and legitimate knowledge was ultimately defined by large software publishers who, within the framework of private training structures, granted certifications, which played an important role in the recomposition of professional hierarchies:

> This certification system is based on – and accelerates – an IT workforce selection process based on standardized, transferable skills, applicable across various spaces and countries, granted a sufficiently buoyant market, and this, concomitant with the exclusion of holders of skills specific to a product or a company always threatened with disappearing, with a feeble possibility of mobility (Stevens 2010, p. 49).

3.3.2. *Lean management, an exacerbation of the tensions associated with the standardizing ambition*

The ambition to standardize work processes, long supported by total quality management, was later relayed by managerial projects inspired in lean management. After sustained interest in Toyotism as a productive model in the 1980s and 1990s, lean[4] received renewed attention after the mid-2000s. Reaching beyond the mass-produced industry where its precepts were conceived[5], lean presented itself as "a general, all-terrain method, capable of unveiling the squandering of resources which carry costs without really adding value for the customer" (Ughetto 2012, p. 150). Its promise was to produce as much, or even more, with less. In this sense, it immediately appeared as a project for rationalizing work activities, for removing or continually reducing time deemed unproductive and, more generally, anything failing to bring added value to the product from the customer's point of view. The implication for workers was obvious (and this fueled the main criticisms to *lean*): work intensification generated a feeling of deterioration in working conditions, "a compaction of time which put operators in relentless tension with production" (ibid., p. 158). The moments for recovery disappeared, despite being needed for reflective engagement in the activity, essential for cooperating, projecting themselves into the rest of the activity, adjusting operating methods, etc.

But rationalization also resulted in an increase of responsibility at work, attention being placed on solving problems and dysfunctions typical of ordinary activity, bringing managers close to the operational and practical challenges inherent in the productive activity. Staff involvement was promoted in order to engage those close

---

4 We use the conventions adopted by Ughetto (2012), who made a distinction between lean production (the model designed from the automobile industry, as in the Toyota production system), lean manufacturing (model aimed at extending to all industrial sectors) and lean management or lean (recent managerial mode, aiming to be implemented across all occupational activities, and services in particular).

5 Lean has been applied in most industrial sectors. It also served as a model in services, hospitals and justice (Vigour 2015).

to the activity in the design and implementation of the necessary changes[6]. This form of professionalization is, to say the least, ambivalent. The continuous improvement approach, characteristic of lean production, was instrumented by the accurate measurement of yields and operating times, and strictly oriented toward improving productive performance. In other words, an employees' ability to modify their activity and their work context was only permitted provided that it put an end to resource squandering. While finding themselves invested with responsibilities the scientific organization of work had deprived them from to the benefit of the hierarchy, employees were called to judge on quality and to intervene in the production process, only in the respect and for the benefit of the operating standards set by others. Analogous to quality management (see above), the lean model was characterized by a tension, inherent in it, between the rationalization of processes (activity scheduling, task standardization, etc.), on the one hand, and the call for autonomous resolution of productive problems close to the activity and relief from the constraint of accountability (of reporting, purely bureaucratic hierarchical validation, etc.), on the other hand. The way in which this tension is solved depends on work scenarios. This invites us to understand lean in a plural way, in function of the productive spaces and professional contexts where we seek to implement it:

> we are confronted with a tightly closed lean production for the work activity which nonetheless offers certain openings. Everything will depend on the social processes for the construction of lean *in situ* (Ughetto 2012, p. 162).

The weight given to the measurement is also a decisive element in the interpretation of lean.

Pardi (2007) showed that within the same multinational company, lean modulations may vary depending on the institution, with a stark contrast between bottom-up and top-down operating modes. In the first case (bottom-up), as in the British establishment studied, proposals for adjustment rise from the operators to the facilitators, who choose to speak about them to their superiors (or not) and these, in turn, are likely to adjust the standard. In the second case (top-down), which seems more typically French, adjustments are initiated by the hierarchy and are the result of a discussion between facilitators, and often without the operators. The production system in place does not correspond to the original model, but is the result of hybridization with certain societal characteristics of labor relations, such as the antagonistic labor management relationship model from the United Kingdom (Pardi 2005).

---

6 We will later return to the specific analysis of the collective discussion times contemplated in the lean model (see section 5.2).

3.4. New Public Management: work standardization for public employees?

The expression "New Public Management" (NPM) designates new managerial practices having developed in the public sector over the past 40 years[7], with the goal of introducing management methods used in private companies. These new practices were backed by the development of a set of standards and tools aimed at the standardization of professional practices through increased managerial control, especially within the framework of contractualization programs. The hospital was described and studied as the exemplary sector for these transformations and their effects on professional groups in France (Belorgey 2010), Europe (Kirkpatrick et al. 2009; Kuhlmannn et al. 2009) and the United States (Scott et al. 2000). The desire to control health expenditure resulted in the establishment of activity-based pricing and the greater presence of (nonmedical) management professionals in the governance of health institutions (Feroni and Kober-Smith 2005). The creation of managerial positions not reserved exclusively for professionals has also been observed in the justice sector (Dugué and Mallochet 2008; Vigour 2008).

New Public Management also resulted in the establishment of instruments for piloting and evaluating the activity, based on results or performance indicators, eventually affecting the definition of work and thereby transforming what those instruments were supposed to measure. The deployment of these accounting techniques was combined with the dissemination of "best practices" standards (Harrison 1998), intended to promote transparency and service quality. By this means, NPM ended up imposing action conventions and normative references, failing to match the way in which professionals understand their missions and their activities. In the case of hospital emergency departments, Belorgey (2011) demonstrated how the desire to increase the health care rate (underlying the pressure to reduce the patients' waiting time) translated into medical practices which degraded the quality of care, and ended in a comeback from patients to request still new care practices.

However, we should not oppose these two conceptions of activity (managerial and professional) too systematically. If we consider the case of the French National Police, institutional change was accompanied by the affirmation of a "culture of results" and the establishment (or rather the strengthening) of performance indicators with the goal of "piloting" public action. This injunction to "produce numbers" certainly has an impact on certain professional conceptions of police activity. However, as Boussard (2010, pp. 107–119) cleverly pointed out, these new managerial standards simultaneously coincided with and reinforced a traditional

7 Even older in Great Britain. Among the numerous references regarding the British case, see, in particular, Kirkpatrick et al. (2005).

definition of activity, focusing on the fight against crime and delinquency. It is not so much a question of opposition and confrontation but of concealment and segmentation effects: an obscuration, even a form of disqualification, of non-repressive activities, falling under the scope of prevention or assistance to the victims; increasing segmentation of the professional group, accentuating internal divisions, depending on the distance from the "professional ideal" reinforced by the "culture of results". de Maillard and Savage (2017) noticed diverse reactions to the neo-managerial reforms of the British police, which were manifested through the heterogeneous appropriation of performance indicators.

Backed by the goal of justice efficiency, the "real-time" scheme for processing criminal cases, implemented in the 2000s, seems emblematic of the import of NPM into the legal world. This scheme prompted police services to quickly communicate the offenses observed to the public prosecutor's office, who had to give an immediate direction to the case. Echoing what was observed for the police, Bastard and Mouhanna (2007) revealed that this scheme was not imposed bottom-down, but had found a favorable reception among the actors of the repressive scheme. Far from imposing a rigid functioning, the scheme was subject to local adaptation matching the specificities of the high courts considered. Its effects on the work of prosecutors can in no way be reduced to an alternative between conformity and resistance. Bastard and Mouhanna described a significant modification of this type of work, which the actors praised as a new métier combining tension and exaltation. While "real-time processing" generated significant constraints and a sustained work pace, it was also experienced as an opportunity to face reality head-on for those who adopted the repressive vision embodied by the scheme. Here again, the reactions were diverse and varied depending on the different segments of the professional group at the prosecution office.

It was also with the ambition of strengthening the effectiveness of its interventions that the labor inspection body was subject to successive reforms after 2006. These reforms were extensive, affecting both the organization of labor inspection (territorial structure, relationship with "users", tendency toward the specialization of agents, etc.) and employment conditions (integration of labor inspectors into the body of inspectors, emergence of a local management métier with "control unit managers"), strengthening of investigation and coercion powers (Tiano 2020). They also resulted in a reinforcement of the control over labor inspectors and the national piloting of intervention priorities, featuring a "labor policy" based on an information system which included figure indicators to monitor and report on the activity of labor inspection services. We proposed interpreting these transformations as a major recomposition of the "cognitive legitimacy" of inspection, of its capacity to make its actions and purposes understandable and self-evident. Maintaining this legitimacy requires increased communication and cooperation with other actors, which exacerbates the traditional tension between control and pedagogy (Dodier 1989, pp. 281–306) by multiplying the situations where agents must compromise, therefore affecting the exercise of their professional autonomy (Mias 2015).

What we homogenize under the name New Public Management actually spans a vast doctrinal diversity (Bezes 2005). The reforms inspired by it are often incomplete, maintaining certain traditional bureaucratic mechanisms unchanged. They disrupt the professional practices of groups whose consistency and legitimacy are not the same (Bezes and Demazière 2011), which may have contrasting effects depending on the case. All in all, the implementation of NPM cannot be interpreted as a uniform movement of deprofessionalization or loss of autonomy. According to Newman (2001), while this implementation has resulted in a standardization of procedures and control of results, it has also boosted the capacities for local initiative, as well as the expertise and the self-regulation capacities of professionals. In France, the "modernization" of public services did not systematically reduce the sources of autonomy at work. Demazière (2010, pp. 51–61) explored the way in which the evolution of the postal service (from the administration by the PTT to the public company, with the rise in profitability requirements and the affirmation of a commercial logic), affected the activity and working conditions of postmen. Recruitment procedures evolved, giving priority to the aptitude for commercial activities (with the subsequent adherence to the company's new orientations) over the former morality check. Procedural standards also modified distribution conditions (namely, the itinerary), reinforcing mandatory tasks and assessments by the hierarchy. But the organizational core of distribution work remained in the hands of the postmen, both collectively (the collective regulates the allocation of itineraries) and individually (the adjustment of professional practice and the actual implementation of said itineraries). Procedures were regularly violated while arrangements to meet the rule were justified by a certain conception of the quality of service provided:

> These arrangements were invariably justified by the desire to serve the customer and reflect a conception of a job well done, of the quality of service, which differed from the management's position. Above all, these practices were not individual rebellions. They resulted from exchanges among postmen and crystallized practical and collective knowledge inscribed in – and writing out – the history of the métier. This experiential knowledge represented a kind of informal and implicit ethics, holding decisive weight, which played a more central role than the code of formal procedures, in the regulation of practices (Demazière 2010, p. 56).

The possibilities for managerial control were harnessed, while the postmen retained their capacity to master the service relationship with users.

3.5. Experimenting with new forms of labor governance

At the turn of the 2010s, as psychosocial risks made their appearance on the agenda, a normative proposal for experimenting with new forms of labor governance was expressed, promoting spaces for discussion regarding work. It quickly experienced a first institutionalization within the framework of the French National Inter-professional Agreement of 13 June, 2013 aiming at "a policy for improving the quality of life at work and professional equality". Considering that "giving employees the possibility of expressing themselves regarding their work, the quality of the goods and services they produce, the conditions for performing such work and work efficiency, is one of the elements promoting their perception of the quality of life at work and the meaning given to work", Article 12 of the agreement ("Encouraging and promoting the expression of employees regarding work"), encouraged the setting up of spaces for discussion in the form of "working groups of employees from a homogeneous entity producing or providing a service". These could "be organized in the presence of a professional referent or facilitator responsible for leading the group, framing their expression and including time in the presence of their hierarchy". The agreement also established that the reporting resulting from those exchanges should provide the employer with "elements for reflection on, on the one hand, possible developments in the organization of work geared toward greater autonomy, and on the other hand, reflection on the role and means for enhancing management". Beyond the "quality of life at work", those discussion spaces were designed as places targeting the transformation of labor governance. The implementation of these discussion spaces was presented as an experiment, which would require making a balance 3 years after the signing of the agreement.

This normative proposal was part of a set of largely independent initiatives, whose goal was to improve productive performance, rather than well-being at work. Regardless of their origin and the argumentative rhetoric accompanying them, such proposals for new forms of labor governance are relevant to our analysis insofar as they primarily aim to strengthen the capacity of workers to define the content of their activity and to modify the organizational and relational environment where their contribution unfolds. Those proposals outlined a path toward professionalization which Wittorski (2009, pp. 781–793) had described as "the logic of reflection for action". They offered the groups thus formed the possibility of defining new ways of doing things, of improving quality or efficiency at work, and of putting their skills (deemed legitimate) into words, through the formalization and collective validation of professional knowledge.

3.5.1. *The "liberated" company: a disruptive proposal?*

Proposals in favor of employee autonomy have sometimes taken a somewhat confusing turn. If we admit that management action is generally far more concerned with direct productivity than the rehabilitation of initiative at work, the thesis of the liberated company, promoted by a few charismatic entrepreneurs and relayed in various publications[8], stands out in the business landscape, to say the least. The proposal is relatively simple. It purports reducing hierarchical management to what is strictly necessary and entrusting autonomous work collectives with the freedom and responsibility to make all the decisions they deem appropriate as to productive organization, the division of labor, even remuneration and bonuses. The business manager is summoned to become a leader who must decide on the main strategic orientations and set the course. By precisely explaining their "vision", business managers must create the conditions that will encourage everyone's commitment to work. However, they should not waste time producing precise standards and monitoring compliance with them. Two main principles are at work here. The first is that the managerial control requested in traditional organizations is particularly costly, not only because the resources involved in ensuring this control, but also because it represents a straitjacket stifling the creativity of employees, maintaining generalized distrust and ultimately leading to their disengagement. As a corollary to the first, the second principle postulates that employees spontaneously seek to do a good job. If we give them the opportunity to do so, they will be a force for proposals, initiative and value creation. As Ughetto (2018) suggested, this principle recalls the opposition that Douglas MacGregor drew in the early 1960s between "Theory X" and "Theory Y". The first, at the basis of the scientific organization of work, postulates that individuals spontaneously seek to minimize effort and avoid work. On the contrary, "Theory Y" calls for associating individuals with the goals of the organization, letting them arrange themselves as it best suits them and giving them the opportunity to learn at work. It is under these conditions that they will fully commit to their work. We could also mention the connection with the sociotechnical approach developed in the 1970s (Gilbert et al. 2017). The "liberated company", apparently disruptive, makes part of a long history of proposals in favor of participation at work, such as semi-autonomous groups or quality circles, whose limitations were quickly detected (Linhart 1994; Martin 1994). Participatory schemes lend themselves to diversified uses (Borzeix et al. 2015), as an instrument to support organizational change, as in the case of the French Ministry of Finance studied by Spire (2015), as a means for asserting a form of authorship over our

8 The pioneering work of Peters (1992) inspired those publications, the most commented on being those by Getz and Carney (2012) and Laloux (2015).

own work (Dupuy 2015), or as "contributory participation" (Zask 2009), to support productive activity.

There is still a lack of rigorous empirical work to understand the effects of this type of transformation, labeled as the "liberated company". At best, we can see that while certain "liberated" companies experienced the promised economic and social success, other leaders quickly took a step back, in the face of the numerous difficulties raised by that shift. Let us also mention that the organizational formulas found locally may vary considerably and that heterogeneity predominates among "liberated" companies. Pitfalls could thus be identified. Studying three cases of companies of this type, Gilbert et al. (2017) highlighted the many common features with the participatory schemes of the 1970s–1980s, while showing that participation is now asserting itself in the daily life of work, and no longer in meetings (interludes in productive activity). While certain pitfalls of participation at work seem to have been overcome, other difficulties have appeared: a loss of reference points caused by the disappearance of the old without offering anything new; peer regulation (struggling to be put in place); managerial withdrawal from conflict regulation; contrasting logics of commitment which can lead to weariness among those most involved. We could add that "liberated companies" generate forms of exclusion (of those who "do not play the game") and, conversely, the over-commitment of the most involved.

3.5.2. *What regulation of autonomy at work? The example of lean*

Access to autonomy has a positive connotation. But this valorization tends to ignore the conditions in which such autonomy can be exercised without drifts or suffering. As numerous works in work psychology have shown, there is a risk of exhaustion in wanting to solve certain problems on our own, taking it all upon ourselves to act properly. However, as Ughetto (2007, 2018) pointed out, it is the activity that often lacks organizational and managerial support. The activity needs to be organized. One of the risks posed by the new labor standards is that they isolate employees in the face of work situations and difficulties.

This lack of supervision and regulation of work which some analysts have detected in the experiences of "liberated companies" (see above) was observed in the organizations which opted for lean management. While this manifests as an ambition for rationalization and standardization (see above), operationally, it translates into a call for employee autonomy and empowerment in relation to the organization of their work. It provides for the creation of autonomous and versatile production cells based on work standards (*jidoka*), with the goal of encouraging all company employees to improve the quality of the products and services sold ex ante.

During short meetings (a few minutes), the discussion focuses on the difficulties found by employees so as to identify operational challenges as quickly as possible and to readjust standards. These participatory approaches appeal to employees on a daily basis. The effectiveness of such a scheme relies on the long-term commitment of employees to the participatory approach. The challenge is therefore to ensure the favorable conditions and the necessary means to grant such commitment, especially since the participation of employees is far from assured. For fear of control, they may have an interest in keeping quiet about their difficulties in order to maintain their leeway, even if this is precarious and imperfect. Furthermore, autonomous production cells do not spontaneously generate cooperation, and even tend to reinforce peer pressure, recalling the paradoxical effects of "open participation" schemes analyzed by Borzeix and Linhart (1988). They can act as a source of tension within work groups. These tensions have various origins, such as the way in which work is distributed within the team or the degree of investment of each person in the collective activity. Finally, attitudes toward the schemes amply vary, as Matt Vidal pointed out with regard to lean: "there appears to be a normal distribution of worker attitudes toward increased employee involvement, where a relatively small proportion of workers are enthusiastic about the opportunity to participate in substantive decision-making and problem-solving, a relatively small proportion are staunchly opposed and the majority somewhere in the middle. For most workers predictability at work, good interpersonal relations, effective communication from management, and job security appear to be more important in providing a positive work context than opportunities for substantive participation in decision-making and problem-solving" (2007, p. 272).

Beyond these difficulties inherent in participatory schemes, we wonder whether, in their concrete functioning, these autonomous production cells offer sufficient leeway letting workers redefine the regulation of their work, individually and collectively. The key seems to lie in linking these team discussions with the management line. In such a configuration, local supervisors are invited to organize discussions, encouraging the staff to identify problems, reflect upon them and propose countermeasures to improve the production system. They are invited to stay close to the productive activity and to the problems which may arise. But at the same time, these managers are often asked to grant the deployment of tools that operators do not support, have great difficulty in managing the problems identified and even reporting them. Managers are placed in a most uncomfortable position: compelled to rely on reporting tools, they are immediately suspected of exercising control over their teams; they listen to employees' complaints regarding their working conditions, without having the means to provide a response. Hardly do they have the legitimacy or the necessary autonomy to support the activity of their subordinates. Fieldwork carried out by Bertrand and Stimec in three companies having implemented lean management principles focused on this sometimes untenable position of middle management:

> Permanently confronted with the deficiencies of the organization but unable to change anything, required to relay contradictory injunctions coming from all sides and discredited by the hierarchy, middle management finds itself in an untenable situation (2011, p. 141).

Above all, Bertrand and Stimec exposed the diversity of applications of lean management, inviting us not to generalize the issue. According to them, diversity is due to the difficulties in applying the lean approach in a sustainable and comprehensive manner, particularly at the level of participatory schemes. A first quantitative analysis (Beauvallet and Houy 2009) showed that the operational variation of lean was often implemented without setting up formal spaces for discussion. This reflects a tendency to consider *lean* as an instrument for industrial performance (the fight against squandering and downtimes, and the transfer of best practices), without acknowledging the organizational context and the effects on work activity and health (Longoni et al. 2013). However, practices inspired in lean are changing working methods and transforming métiers through the redefinition of the means and criteria to assess high-quality work. In the service activities, they can severely alter the relational dimension of work in contact with users or customers, which is nonetheless the dimension which generally engages the most employees. When subject to partial application (focusing on immediate productivity gains), they are also likely to cause dissatisfaction and disengagement (Bouville and Alis 2014).

3.5.3. *The problematic engineering of discussion spaces*

These observations stress the importance of discussion spaces intended to bring managers closer to the working context of their employees, to analyze the activity, to organize an in-depth discussion on the conditions of its realization, and to support the staff in the difficulties they face on a daily basis. Designed in this way, discussion spaces cannot be reduced to formal meetings where information is exchanged and solutions quickly found to satisfy the customer, but should help and empower employees to decide on the problems encountered and to hierarchize priorities. Various experiments have been carried out in recent years, under the watchful eye of ergonomics or management sciences, often drawing on the thoughts of researchers in psychology. They have recently been the subject of various analyses which study their scope and challenges.

Among them, the intervention of Yves Clot's team at the Renault factory in Flins presents many features of this paradigmatic case. A recent work (Bonnefond 2019) finely recreates its stages theoretical foundations. The organizational context, marked by a form of lean management promoting the optimization of productive schedules and the strong recourse to temporary workers, immediately appears as a hostile environment for any discussion regarding work: first, we must start by overcoming the idea that discussion times are supposedly unproductive. In addition to this context there is a degraded social climate and a disagreement between management and unions on the diagnosis of the situation in terms of health at work, as well as the repeated and shared experience of "useless speech". Even though the operators speak, report and propose, this ends up being relevant; tired of reporting problems, they opt for silence and strive to do their best, often at the cost of their health. By relying on filmed work sequences and a cross-self-confrontation scheme (already tested in other interventions in activity), psychologists initiated a transformation process leading to the choice of "reference operators" called upon to record the difficulties encountered at each work station and the possible solutions devised. The investigative work of these referents feeds a "single list of problems" which, shared and publicized, serves both as a tool for monitoring the treatment of said problems and as a lever for appealing to the managerial line regarding the difficulties and dilemmas still requiring a solution. "The elected reference operators embody a new professional authority, legitimized by the other operators. They act as a lock, an institutional force reminding the work collective that they have to bring the organization and the unions down to real work" (Bonnefond et al. 2019, p. 76). Beyond this challenge to bring the organization closer to the concrete demands of work, this experiment opens spaces for controversy – more than discussion – on the subject of work. It makes it possible for the contradictory plurality of representations on a job well done to be expressed first between professionals, then between them and the hierarchical line (Clot 2010). This confrontation between the different representations of work quality nourishes what Clot has called the "professional genre", for the benefit of both productive efficiency and health. Such a confrontation reveals a fundamental difference with the discussion times of lean management, which ergonomists Bourgeois and Gonon (2010) have assimilated to a "new doctrine of efficiency".

Whether this takes the form of a controversy around the challenges of work quality (or not), the discussion on work promoted by the experiments discussed so far differs in principle from any form of top-down communication and piloting tools. Dialogical in nature, this discussion is accompanied by particular requirements for the workers who engage in it (Detchessahar 2012). To begin with, it involves making our own representation of work a public issue, exposed to the gaze of others. It induces a formal notice to argue and justify this opinion, and to support the controversies the discussion may generate. It also has a normative goal, which requires having a grip on work rules. While this discussion does not go without a (cognitive and social) cost for those who engage in it, it also requires being organized and sustaining an effort, which in turn presupposes certain managerial and organizational conditions being met. Echoing what we discussed on the subject of lean[9], one of the main challenges concerns the place of managers (Detchessahar 2019). Facilitators must also be an active stakeholder to prevent the discussion from becoming locked into a verbal exchange without any influence on the real work and its regulation. This assumes that they share a common professionalism with the group of workers, in the sense that they know the métier and the activities of the staff they supervise, they have a certain appetite for confronting practical problems in the field and their resolution, and they benefit from relative stability in their position. As we can see, these conditions are far from being facilitated by the evolution of human resources management, which promotes mobility and the diversification of professional experiences and tends to prioritize managerial skills over technical skills in the careers of executives. Furthermore, as Ughetto (2007) or Detchessahar and Minguet (2012, pp. 233–247) have shown, certain characteristics of contemporary organizations prevent managers from fully exerting their role as managers, supporting productive activity. Monopolized by the maintenance and feeding of "management machines" (Girin 1983) and torn between a disparate set of meetings (project groups, information meetings, training, etc.), they are kept at a distance from the work situations of their collaborators. "Unblocking" management means giving back the time, means and incentives for them to take charge of local regulatory work by refocusing managerial activity on the animation and support of work collectives.

But the effectiveness of discussion spaces is not only due to the position of managers in the discussion. It also depends on the place of the latter in the organization, in other words, on the leeway managers have. It requires the delegation of authority, which often involves rebuilding the local decision-making power of the local supervisor to solve the problems identified in the discussion and propose responses and support. This also assumes that central management tolerates local arrangements with the rules they edict, which is not obvious, so powerful are the tendencies toward standardization and centralization. A singular vision of managerial authority is therefore invited into these experiments, which some researchers associate with the concept of subsidiarity (Melé 2005; Petit et al. 2011; Rocha et al. 2017).

3.6. Conclusion

This chapter was organized around major axes of organizational transformation along which contemporary managerial strategies are positioned. It explored the

9 But this also applies to so-called "liberated" companies, which seem not to provide the conditions for such a dialogue around the question of work (Detchessahar 2019).

effects on autonomy at work and the responsibility of workers in order to question the way in which these new standards and requirements recompose professionalization dynamics. From this emerges a set of tensions which structure the contemporary work experience: between autonomy and intensification; between accountability, exposure to others and engagement in cooperation; between standardization and margins for local adaptation; between increased managerial control and reaffirmation of professionalism; between opening spaces for discussion at work and maintaining a unilateral representation of productive efficiency; between the call for collective deliberation and individual responsibilization; etc. Under these conditions, the ways of transforming work are multiple and their effects necessarily ambivalent. We are undeniably witnessing a profound recomposition of professional autonomy and responsibility, which cannot be reduced to a diagnosis in terms of increase or reduction.

The decline in professional autonomy, as defined in the introduction, is incontestable. The capacity of groups of workers to define what work should be and to control the standards that govern their activities is eroding. But as we have shown, the possibilities for resistance, circumvention and local regulation have not been definitively eradicated. The direction of developments is far from univocal. The same goes for learning conditions, the support for professionalization. While the contemporary work environment appears in many respects restrictive and inhibiting, it also increases the opportunities to learn new things, as well as the incentives to engage reflexively in the work activity, in contact with and under the gaze of/in discussion with peers.

We are also witnessing a redefinition and extension of workers' responsibilities. As Lallement and Zimmermann (2019) have suggested, in order to measure its scope, it is appropriate to distinguish different components of responsibility, which are experiencing contradictory developments. In that sense, responsibility is "both a *state* (I am responsible for such and such a person), an *ability* (I have resources, such as will power or discernment) and an *obligation* (I must take responsibility for my actions)". The first component refers to the register of care: to be responsible for someone or something is to be in charge of them. According to the second component, to be responsible, we have to possess the capacity to assume the attribution of an action and its consequences. The last component refers to the obligation to account for our actions. Although the injunctions to "account for" and to "take charge" have been strongly affirmed across the managerial and organizational innovations analyzed in this chapter, rarely have they been accompanied by a consolidation of resources to "be capable of" something. While being held more accountable for their work, workers still often struggle to access the necessary resources to act in compliance with the conditions of their activity.

3.7. References

Amira, S. and Ast, D. (2014). Des risques professionnels contrastés selon les métiers. *DARES Analyses*, 39, 17.

Appay, B. (2005). *La dictature du succès, le paradoxe de l'autonomie contrôlée et de la précarisation*. L'Harmattan, Paris.

Askenazy, P. (2004). *Les désordres du travail*. Seuil, Paris.

Bastard, B. and Mouhanna, C. (2007). *Une justice dans l'urgence. Le traitement en temps réel des affaires pénales*. PUF, Paris.

Beauvallet, G. and Houy, T. (2009). L'adoption des pratiques de gestion lean. Cas des entreprises industrielles françaises. *Revue française de gestion*, 197, 83–106.

Belorgey, N. (2010). *L'hôpital sous pression. Enquête sur le nouveau management public*. La Découverte, Paris.

Belorgey, N. (2011). "Réduire le temps d'attente et de passage aux urgences" : une entreprise de "réforme" d'un service public et ses effets sociaux. *Actes de la recherche en sciences sociales*, 189, 16–33.

Benallah, S. and Domin, J. (2017). Intensité et pénibilités du travail à l'hôpital : quelles évolutions entre 1998 et 2013 ? *Travail et emploi*, 152, 5–31.

Bèque, M. and Mauroux, A. (2017). Quelles sont les évolutions récentes des conditions de travail et des risques psychosociaux ? *Dares Analyses*, 82, 10.

Bernard, S. (2012). *Travail et automatisation des services : la fin des caissières ?* Octarès, Toulouse.

Bertrand, T. and Stimec, A. (2011). Santé au travail : voyage en pays de lean management. *Revue française de gestion*, 214, 127–144.

Bezes, P. (2005). Le renouveau du contrôle des bureaucraties. L'impact du New Public Management. *Informations sociales*, 126, 26–37.

Bezes, P. and Demazière, D. (2011). Introduction au dossier débat : new public management et professions dans l'État : au-delà des oppositions, quelles recompositions. *Sociologie du travail*, 53(3), 294–305.

Boboc, A. and Metzger, J.-L. (2020). Les méthodes agiles et leurs contradictions. *SociologieS* [Online]. Available at: http://journals.openedition.org/sociologies/12471.

Boltanski, L. and Chiapello, È. (1999). *Le nouvel esprit du capitalisme*. Gallimard, Paris.

Bonnefond, J.-Y. (2019). *Agir sur la qualité du travail. L'expérience de Renault-Flins*. Érès, Toulouse.

Bonnefond, J.-Y., Scheller, L., Clot, Y. (2019). Pouvoir agir sur la qualité du travail pour développer la santé et la performance. Une expérience à Renault-Flins. In *Agir sur la santé au travail*, Mias, A. and Wolmark, C. (eds). Octarès, Toulouse.

Borzeix, A. and Linhart, D. (1988). La participation : un clair-obscur. *Sociologie du travail*, 30(1), 37–53.

Borzeix, A., Charles, J., Zimmermann, B. (2015). Réinventer le travail par la participation. Actualité nouvelle d'un vieux débat. *Sociologie du travail*, 57(1), 1–19.

Bourgeois, F. and Gonon, O. (2010). Le lean et l'activité humaine. Quel positionnement de l'ergonomie, convoquée par cette nouvelle doctrine de l'efficacité ? *Activités*, 7(1), 136–142.

Boussard, V. (2010). Les policiers de la sécurité publique et l'obligation de faire du chiffre. In *L'injonction au professionnalisme. Analyses d'une dynamique plurielle*, Boussard, V., Demazière, D., Milburn, P. (eds). Presses Universitaires de Rennes, Rennes.

Bouville, G. and Alis, D. (2014). The effects of lean organizational practices on employees' attitudes and workers' health: Evidence from France. *International Journal of Human Resource Management*, 25(21), 3016–3037.

Campinos-Dubernet, M. and Marquette, C. (1999). Une rationalisation sans norme organisationnelle. La certification ISO 9000. *Sciences de la société*, 46, 83–101.

Carney, B.M. and Getz, I. (2009). *Freedom Inc: Free Your Employees and Let Them Lead Your Business to Higher Productivity, Profits and Growth*. Crown Business, New York.

Castel, P. and Dalgalarrondo, S. (2005). Les dimensions politiques de la rationalisation des pratiques médicales. *Sciences sociales et santé*, 23(4), 5–40.

Clot, Y. (2010). *Le travail à cœur. Pour en finir avec les risques psychosociaux*. La Découverte, Paris.

Cochoy, F., Garel, J.-P., de Terssac, G. (1998). Comment l'écrit travaille l'organisation : le cas des normes ISO 9000. *Revue française de sociologie*, 39(4), 673–699.

Courpasson, D. (2000). *L'action contrainte. Organisations libérales et domination*. PUF, Paris.

Coutant, H. (2018). L'ambivalence du "métier". Métier de l'organisation contre métier de l'ingénieur dans une entreprise aéronautique. *Sociologie du travail*, 60(4) [Online]. Available at: https://journals.openedition.org/sdt/8118.

Coutrot, T. (1998). *L'entreprise néolibérale, nouvelle utopie capitaliste ?* La Découverte, Paris.

Demazière, D. (2010). Les facteurs et les évolutions du service postal. In *L'injonction au professionnalisme. Analyses d'une dynamique plurielle*, Boussard, V., Demazière, D., Milburn, P. (eds). Presses Universitaires de Rennes, Rennes.

Detchessahar, M. (2013). Faire face aux RPS : quelques éléments d'un management par la discussion. *Négociations*, 19, 57–80.

Detchessahar, M. (2019). *L'entreprise délibérée : refonder le management par le dialogue*. Nouvelle Cité, Bruyères-le-Châtel.

Detchessahar, M. and Minguet, G. (2012). Santé au travail : l'enjeu du management. In *Risques du travail, la santé négociée*, Courtet, C. and Gollac, M. (eds). La Découverte, Paris.

Dodier, N. (1989). Le travail d'accommodation des inspecteurs du travail en matière de sécurité. In *Justesse et justice dans le travail*, Boltanski, L. and Thévenot, L. (eds). PUF, Paris.

Dubar, C., Tripier, P., Boussard, V. (2015). *Sociologie des professions*. Armand Colin, Paris.

Dugué, E. and Malochet, G. (2008). Une légitimité sans assise. Les directeurs de la PJJ entre dilemmes professionnels et contraintes institutionnelles. In *Action publique et légitimité professionnelle*, Le Bianic, T. and Vion, A. (eds). LGDJ, Paris.

Dupuy, C. (2015). Participer pour dépasser la condition salariale : les journalistes en quête d'indépendance. *Sociologie du travail*, 57(1), 61–80.

Eurofound (2021). *Working Conditions and Sustainable Work: An Analysis Using the Job Quality Framework, Challenges and Prospects in the EU Series*. Publications Office of the European Union, Luxembourg.

Evetts, J. (2003). The sociological analysis of professionalism: Occupational change in the modern world. *International Sociology*, 18(2), 395–415.

Evetts, J. (2011). A new professionalism? Challenges and opportunities. *Current Sociology*, 59(4), 406–422.

Faulconbridge, J. and Muzio, D. (2008). Organizational professionalism in globalizing law firms. *Work, Employment and Society*, 22(1), 7–25.

Feroni, I. and Kober-Smith, A. (2005). La professionnalisation des cadres infirmiers : l'effet de l'action publique en France et en Grande-Bretagne. *Revue française de sociologie*, 46(3), 469–494.

Flamant, N. (2002). "On ne sait plus qui est le chef". Travailler et diriger dans l'industrie spatiale. *Terrain*, 39, 109–120.

Ford, R.C. and Randolph, W.A. (1992). Cross-functional structures: A review and integration of matrix organization and project management. *Journal of Management*, 18(2), 267–294.

Fournier, V. (1999). The appeal to 'professionalism' as a disciplinary mechanism. *The Sociological Review*, 47(2), 280–307.

Freidson, E. (2001). *Professionalism, The Third Logic*. University of Chicago Press, Chicago.

Gadrey, J. (1994). La modernisation des services professionnels. Rationalisation industrielle ou rationalisation professionnelle ? *Revue française de sociologie*, 35(2), 163–195.

Gilbert, P., Teglborg, A.-C., Raulet-Croset, N. (2017). L'entreprise libérée, innovation radicale ou simple avatar du management participatif ? *Gérer et comprendre*, 127, 38–49.

Girin, J. (1983). Les machines de gestion. In *Langage, Organisations, Situations et agencements*, Girin, J. (ed.). Presses de l'Université de Laval, Quebec.

Gollac, M. and Volkoff, S. (1996). Citius, altius, fortius. L'intensification du travail. *Actes de la recherche en sciences sociales*, 114, 54–67.

Grey, C. (1997). Management as a technical practice: Professionalization or responsibilization? *Systems Practice*, 10(6), 703–725.

Harrison, S. (1998). The politics of evidence-based medicine in the United Kingdom. *Policy and Politics*, 26(1), 15–31.

Hodgson, D. (2002). Disciplining the professional: The case of project management. *Journal of Management Studies*, 39(6), 803–821.

Hodgson, D. (2005). 'Putting on a professional performance': Performativity, subversion and project management. *Organization*, 12(1), 51–68.

Hodgson, D. and Briand, L. (2013). Controlling the uncontrollable: 'Agile' teams and illusions of autonomy in creative work. *Work, Employment and Society*, 27(2), 308–325.

Inan, C. (2014). Les facteurs de risques psychosociaux en France et en Europe. *Dares Analyses*, 100, 7.

Kipping, M., Kirkpatrick, I., Muzio, D. (2006). Overly controlled or out of control? Management consultants and the new corporate professionalism. In *Production Values: Futures for Professionalism*, Craig, J. (ed.). Demos, London.

Kirkpatrick, I., Ackroyd, S., Walker, R. (2005). *The New Managerialism and Public Service Professions*. Palgrave, London.

Kirkpatrick, I., Kragh Jespersen, P., Dent, M., Neogy, I. (2009). Medicine and management in a comparative perspective: The cases of England and Denmark. *Sociology of Health and Illness*, 31(5), 642–658.

Knights, D. and McCabe, D. (2000). 'Ain't misbehavin'? Opportunities for resistance under new forms of 'quality' management. *Sociology*, 34(3), 421–436.

Kuhlmann, E. and Saks, M. (2008). *Rethinking Professional Governance: International Directions in Health Care*. The Policy Press, Bristol.

Kuhlmann, E., Allsop, J., Saks, M. (2009). Professional governance and public control: A comparison of healthcare in the United Kingdom and Germany. *Current Sociology*, 57(4), 511–528.

Lallement, M. (2010). *Le travail sous tensions*. Éditions Sciences Humaines, Auxerre.

Lallement, M. and Zimmermann, B. (2019). Tous responsables ? Transformations du travail, métamorphoses de la responsabilité. *Sociologie du travail*, 61(2) [Online]. Available at: https://journals.openedition.org/sdt/17981.

Laloux, F. (2015). *Reinventing organizations. Vers des communautés de travail inspirées*. Diateino, Paris.

Linhart, D. (1994). *La modernisation des entreprises*. La Découverte, Paris.

Longoni, A., Pagell, M., Johnston D., Veltri, A. (2013). When does lean hurt? – An exploration of lean practices and worker health and safety outcomes. *International Journal of Production Research*, 51(11), 3300–3320.

Lorenz, E. and Valeyre, A. (2005). Les formes d'organisation du travail dans les pays de l'Union européenne. *Travail et emploi*, 102, 91–105.

de Maillard, J., Savage, S. (2017). Les détectives dans la cage de fer néo-managériale ? Une analyse de deux polices britanniques. *Sociologie du travail*, 59(4), 1–22.

Martin, D. (1994). *Démocratie industrielle. La participation directe dans les entreprises*. PUF, Paris.

McArdle, L., Rowlinson, M., Procter, S., Hassard, J., Forrester, P. (1995). Total quality management and participation: Employee empowerment, or the enhancement of exploitation? In *Making Quality Critical: New Perspectives on Organizational Change*, Wilkinson, A. and Willmott, H. (eds). Routledge, London.

Melé, D. (2005). Exploring the principle of subsidiarity in organisational forms. *Journal of Business Ethics*, 60(3), 293–305.

Mias, A. (2015). Autonomie des agents et légitimité de l'inspection du travail. *La nouvelle revue du travail*, 7 [Online]. Available at: https://journals.openedition.org/nrt/2314.

Midler, C. (1993). *L'auto qui n'existait pas. Management de projets et transformation de l'entreprise*. InterÉditions, Paris.

Muzio, D., Hodgson, D., Faulconbridge, J., Beaverstock, J., Hall, S. (2011). Towards corporate professionalization: The case of project management, management consultancy and executive search. *Current Sociology*, 59(4), 443–464.

Newman, J. (2001). *Modernising Governance. New Labour, Policy and Society*. Sage, London.

Pardi, T. (2005). Crise, effets de trajectoire et dynamiques sociales dans l'évolution de Toyota Motor Manufacturing UK. *Sociologie du travail*, 47(2), 188–204.

Pardi, T. (2007). Redefining the Toyota production system: The European side of the story. *New Technology, Work and Employment*, 2(1), 2–20.

Peters, T. (1992). *Liberation Management: Necessary Disorganization for the Nanosecond Nineties*. Alfred A. Knopf, New York.

Petit, J., Dugué, B., Daniellou, F. (2011). L'intervention ergonomique sur les risques psychosociaux dans les organisations : enjeux théoriques et méthodologiques. *Le travail humain*, 74(4), 391–409.

Rocha, R., Mollo, V., Daniellou, F. (2017). Le débat sur le travail fondé sur la subsidiarité : un outil pour développer un environnement capacitant. *Activités*, 14(2) [Online]. Available at: https://journals.openedition.org/activites/2999.

Rot, G. (1998). Autocontrôle, traçabilité, responsabilité. *Sociologie du travail*, 40(1), 5–20.

Scott, W.R., Ruef, M., Mendel, P., Caronna, C.A. (2000). *Institutional Change and Healthcare Organizations: From Professional Dominance to Managed Care*. University of Chicago Press, Chicago.

Segrestin, D. (2004). *Les chantiers du manager*. Armand Colin, Paris.

Spire, A. (2015). Les ambivalences de la démarche participative dans l'administration. Le cas de la fusion au sein de la Direction générale des finances publiques (2007–2012). *Sociologie du travail*, 57(1), 20–38.

Stevens, H. (2010). Les informaticiens de R&D et les mutations du marché informatique. In *L'injonction au professionnalisme. Analyses d'une dynamique plurielle*, Boussard, V., Demazière, D., Milburn, P. (eds). Presses Universitaires de Rennes, Rennes.

Tiano, V. (2020). Le nouveau visage de l'inspection du travail : 2006–2020 près de quinze ans de réformes incessantes. *Droit social*, 1, 67–76.

Ughetto, P. (2007). *Faire face aux exigences du travail contemporain. Conditions de travail et management*. Éditions de l'Anact, Lyon.

Ughetto, P. (2012). Le lean : pensée et impensé d'une activité sans relâchement. *Activités*, 9(2), 148–167.

Ughetto, P. (2018). *Organiser l'autonomie au travail*. FYP Éditions, Limoges.

Vidal, M. (2007). Lean production, worker empowerment, and job satisfaction: A qualitative analysis and critique. *Critical Sociology*, 33(1–2), 247–278.

Vigour, C. (2008). Ethos et légitimité professionnels à l'épreuve d'une approche managériale : le cas de la justice belge. *Sociologie du travail*, 50(1), 71–90.

Vigour, C. (2015). Professions in policy and knowledge transfer: Adaptations of lean management, and jurisdictional conflict in a reform of the French public service. *International Journal of Sociology*, 45(2), 112–132.

Wilkinson, A. and Willmott, H. (eds) (1995). *Making Quality Critical: New Perspectives on Organizational Change*. Routledge, London.

Wittorski, R. (2009). À propos de la professionnalisation. In *Encyclopédie de l'éducation et de la formation*, Barbier, J.-M., Bourgeois, E., Chapelle, G., Ruano-Borbalan, J.-C. (eds). PUF, Paris.

Zask, J. (2011). *Participer. Essai sur les formes démocratiques de la participation*. Le Bord de l'Eau, Lormont.

4

Struggles Over the Definition of a "Job Well Done"

Valérie BOUSSARD
Printemps, Paris Nanterre University, France

Professionalization "manufactures" professionals by means of training and the transmission of knowledge and know-how. Resulting both from educational institutions and from work and peer experience, it leads to the construction of expertise whose mastery underlies the professionality of those who can invoke it. Within each particular professional group, this expertise is the subject of a set of standards, far beyond the technical knowledge used for performing such activity. Being a professional involves performing a task while respecting rules. The latter allows us to judge upon what constitutes a "job well done". The definition of the right thing to do, both from the viewpoint of the technical gesture and the social dimensions of how to do it (relations with colleagues, customers, etc.), is the cornerstone of professionalism. This delineates the boundaries between professionals and amateurs, lay people, charlatans, etc. It also draws lines of division between professional groups, as well as within the same professional group (between "good" and "bad" professionals), establishing hierarchies within the group. The definition of a job well done may vary within the professional group, dividing it into several segments.

Nevertheless, the criterion for a job well done, understood as the set of standards governing professional work, is to be determined by the professionals themselves. They are determined internally by the members of the professional group and as

Encyclopedia of Professionalization,
coordinated by Didier DEMAZIÈRE and Richard WITTORSKI.

such, fall within its autonomy. They are transmitted to new entrants to the group by means of socialization processes, which make it possible to "manufacture" professionals, while giving rise to sanctions for those who do not respect them. This autonomy does not prevent the definition of a job well done from being directed to audiences, users, the state or other participants in the work activity, from whom the professional group seeks recognition and legitimation.

Over the last 30 years, this model for professionalism as something autonomous, internal to the group, has competed against other definitions of professionalism that characterize it as exogenous to professional groups, defined by the state, the organization and management, or customers and users. In this context, the definition of a job well done can escape professionals and be dictated, even prescribed, by external actors, calling autonomy into question.

As a matter of fact, autonomy is an important challenge for professional groups. The encounter between the indigenous definition of a job well done and exogenous prescriptions creates tensions and struggles. The struggles waged by professionals to defend their definition of a job well done are organized along a continuum, ranging from institutional struggles to pragmatic ones. The form of these struggles depends on the situation of professional groups, in a configuration mixing up the different actors they interact with: state regulators, employers, users, intermediaries, etc. It mainly depends on the way in which groups have been able to organize themselves so as to have their role in society recognized.

This chapter will bring to light the tensions and struggles surrounding a job well done, either to the interior of professional groups, in the relationships among professional groups, or between professional groups and society. It will first show that the definition of a job well done, the cornerstone of professionalism, organizes boundaries and hierarchies between (and to the interior of) professional groups. Second, it will emphasize that professionalism is to be determined by professionals themselves, in the sense that the definitions of a job well done are constructed and transmitted by the group itself. Third, it will present recent developments inviting the intervention of external actors on the definition of a job well done, which, far from the autonomy model, ends up in prescriptions being imposed on professionals. Finally, this chapter will describe the two main forms of struggles (institutional and pragmatic), through which professional groups seek to face challenges to their professionalism.

4.1. The definition of a job well done, the cornerstone of professionalism

The term "professional group" designates "a group of workers performing an activity with the same name, and consequently endowed with social visibility, benefiting from identification and recognition, occupying a differentiated place in the social division of labor, and characterized by symbolic legitimacy" (Demazière and Gadéa 2009). This symbolic legitimacy assumes that the group has been given a place in the division of labor and sometimes the right (exclusive or not) to perform certain activities, in a given space, in exchange for a remuneration. This authorization to practice, or license in the terms of Hughes (1996), is at the core of the monopoly that certain groups manage to grant themselves, within the particular activity territory. In English-speaking countries, we generally speak of "professions" (as distinct from "occupations"), whereas in the French context, we use the term "established or regulated" professions (Dubar et al. 2015). The attribution of this monopoly is the subject of challenges between groups who compete for the recognition of this right to operate. Nevertheless, this attribution may not be formal and legally recognized, but informal, in the symbolic and normative sense (Boussard 2018). In the latter case, even if the profession is not regulated, the group has symbolically – and therefore, practically – set up a fence which brings the professional group into existence by defining the scope of its activity in relation to others. The notion of a "professional group" makes it possible to hold together these two forms of existence (professions and métiers), one around a legal license, the other around a symbolic license. In this text, the term profession or métier will sometimes be used as an equivalent of professional group.

However, this license cannot work on its own. As expressed by E. Hughes, it is accompanied by a mandate, that is to say, a definition of the activity, in the ample sense of the categories under which the professionals' field of action can be conceived. For example, physicians do not content themselves with defining medical practice, but also offer a definition of the nature of health and illness, and actually, of the behaviors that other people and groups should adopt toward their own work. As the *license* brings the profession into existence, the *mandate* is at the origin of professionalism.

Professionalism is therefore based on the rules that the group gives itself, the rules of the art, by means of which it asserts and judges as to the definition of a job well done (section 4.1.1). These rules of the art are bound to ethical and social dimensions, to a self-conduct orientation, the professional ethos, related to the group's philosophy for characterizing its role (section 4.1.2). The definition of a job well done and its corollary, the professional ethos, are the grounds for the boundaries and hierarchies internal and external to a professional group, which in fact leads to a plural definition of a job well done (section 4.1.3).

4.1.1. *The rules of the art*

The mandate concerns the very content of what the profession covers, by delimiting the duties and responsibilities of professionals vis-à-vis the public and specifying the right thing to do. For example, the mandate of teachers defines what should be taught to children and how (didactics, pedagogy). This depends on the discipline taught. By extension, the mandate is a set of rules that defines professional activity and supervises it. On the part of the professionals themselves, it is translated by the development of rituals, which transform the "métier" into an "art" (Hughes 1996, p. 93). While from the public's viewpoint, the techniques used by professionals are a means to an end (healing, educating, etc.); for professionals they almost become an end in themselves: they must be handled with intelligence, finesse, respecting the ways in which things should be done specific to each profession. For example, while the public expects social workers to simply free the persons monitored from their problems, they will apply themselves to doing so by carefully respecting procedures, developing an ability in the way to address people, etc. (Hughes 1996). In the same way, postmen do not consider their métier as plain mail delivery. Developing contact with the inhabitants, either by a simple greeting or small services rendered, is one of the ways to perform that professional activity correctly (Demazière and Mercier 2013).

The most visible dimension of the rules of the art refers to a form of group deontology. Knowledge is not only technical, but is framed by ways of doing things, acceptable to the group, while others are outright unacceptable. In this sense, professional ability or technical perfection are framed by more general rules on how to practice the métier. To return to the postmen example, the technical ability involves knowing how to organize the itinerary to optimize working time, while limiting distribution errors. This ability is acquired through experience. However, it does not exhaust the meaning given to the activity. Performing the job of a postman properly involves respecting other rules, such as those related to the contact with users: finishing an itinerary on time without having spoken to users or having an itinerary where no contact with users is possible do not correspond to the métier's ideal. Succeeding in terms of professional performance is not an end in itself. It is the means for attaining it that count, which are also subject to rules. For example, the success of an intervention, and the mastery of technical gestures and practical tips, is not the only criterion taken into account for judging on the professionalism of a firefighter (Pudal 2016). Firefighters must demonstrate a sense of brotherhood with their colleagues, a spirit of cohesion, a very particular sense of humor and self-mockery, and the ability not to take oneself too seriously. All of these qualities correspond to the informal rules of "firefighter spirit", that is to say, the right way to be a firefighter.

These rules of the art, visible to all contemporary professional groups, recall those which structured the emergence of métier communities (*communautés de métier*) under the old regime, and which are found across all types of professional corporations (guilds, jurandes, brotherhoods, corporations, etc.) in Europe (Sewell 1983; Jullien and Pauly 2016; Bernardi et al. 2020). The métier community used to obtain authorization from the government to practice in a given territory and a particular market segment. The authorization conveyed both the right to practice the métier and to set the rules: the latter established the different métiers and codified them (quality, price, deadline, necessary training, etc.). This made it possible to settle disputes with neighboring professional groups, in competition regarding certain activities and tasks, and who monitored one another, thanks to these regulations, such as tanners and shoemakers (Rivière 2008). On the other hand, each community constituted a métier, conceived as an art. These rules of the art were expressed in a specialized language, with its own jargon. They expressed the qualities required for the work, by defining the criteria for achieving a "masterpiece", a stage which marked the transition to the state of journeyman (*compagnon*). In "sworn" professions, groups united by an oath and endowed with autonomy and a legal personality, like corporations in the 18th century in France, these rules of the art went beyond the mere gestures relating to production. The members of a métier were united under the same "esprit de corps" (Sewell 1980). They shared common feelings, values, worldviews, strongly maintained by a certain number of rites (worship of the patron saint, organization of funerals and weddings, etc.). They also had reciprocal duties related to the vow of fraternity they took, the moment they joined the corporation.

4.1.2. *The professional ethos*

This return via the historical forms of existence of professional groups emphasizes the very presence of an ethical dimension permeating the rules of the art: they include an idea of the duty professionals must conform to, strictly professional acts being overshadowed by a more general orientation in self-conduct. This dimension of a moral ideal, formalized by Weber (1905) through the term *ethos*, helps us to understand that professional groups do not merely propose a definition of the tasks to be performed and the ways of accomplishing them, but more generally, of their role in society. This role is based on a conception of the social object over which the profession intervenes and on a "philosophy" of said intervention (Hughes 1996, p. 67). For example, professions addressing health issues develop a particular representation of illness, those claiming to act on the social order offer a specific representation of delinquency, etc. The professional *ethos* is presented as a system of rationalization of practices and behaviors, relating to the practice of a métier, in connection with this philosophy. It concerns any dimension (epistemic, aesthetic, social, etc.) relating to the practice of a métier (Zarca 2009, pp. 351–352). It is both an orientation for action, in the form of dispositions to perceive and act, and a model for judging your actions and those of colleagues. For example, mathematicians have ways of judging themselves and their colleagues which focus on the reasoning's rigor and the administration of proof. This must be in balance with the intuition manifested, which presupposes putting the concern for demonstration on hold. Proof must also be elegant in the sense of short, concise and minimal. The beauty of the reasoning here opposes its usefulness. Finally, practicing mathematics is associated with pleasure, as something valid for its own sake (Zarca 2009). This set of manners for judging and acting constitutes their professional ethos. It is the foundation of professionalism, understood as the set of rules for defining what a job well done is, or what a true professional is. In this perspective, professional work or a job well done is not so much work carried out according to technical and formal prescriptions, but work which respects ethical and unwritten rules of the art.

4.1.3. *Boundaries and hierarchies*

A job well done goes beyond the simple application of knowledge and techniques. It is about mastering the rules of the art, which have a moral and social dimension and forge a professional ethos for each group. Professionalism, in this ample sense, is the basis for the group's boundaries in relation to users (who are thus reduced to the rank of amateurs) and other professions (those competing or related ones). It also sustains internal relationships within the group by structuring internal hierarchies.

In the first place, the definition of a job well done operates a division of the social world between those who have a professional mandate and those to whom this mandate is addressed, its public, considered as profane. Students who become physicians "pass through the mirror" (Hughes 1958, Chapter 9), have left the "profane" culture, that of the common world, and integrated a professional culture which changes their world view: they now see it through the sick, the illness, the places (the hospital) where patients are taken care of, the tasks they have to accomplish and the career they can tackle. Becoming a police officer also means considering yourself as separate, cut off from the common world, and all the more because this is considered dangerous (Pichonaz 2017). We see that the professional ethos draws a boundary between professionals and those to whom their intervention is addressed.

In the second place, the definition of a job well done at the origin of the boundaries between different professional groups. This delimitation principle between professional groups is articulated around their specific roles and their claim to take charge of problems. Different professional groups effectively compete to take charge of problems and occupy the same activity territory. For example, in the Middle Ages, shoemakers generally obtained the exclusive right to make shoes, but this was regularly challenged – from a legal perspective or factually – by the tanners who supplied the leather and could also claim they made shoes. The same applied to cobblers, in that they were responsible for repairing shoes, and surely knew how to make them (Rivière 2017). On a similar note, different professional groups today claim their ability to plan cities: architects, urban planners, economists, engineers and landscapers (Chadouin 2021). While the concrete and practical knowledge that professional groups develop is important to be recognized as legitimate to occupy an activity territory (to the detriment of competitors), this cannot suffice to ensure lasting control over said activity. It is by means of an abstraction activity conceptually and morally defining the type of problems to be treated and the role of the métier that the latter can ensure the occupation of an activity territory and defend it against the attacks of competitors (Abbott 1988). The production of this abstract knowledge is also what enables the profession to propose a philosophy of its role and of its place in society in relation to other professions. For example, it is through this abstraction process that physicians were able to establish themselves as a prestigious profession. On the one hand, they occupied a place in the social elite of Western societies; and, on the other hand, they excluded other professional groups from any legal medical practice (healers, priests, funeral directors, etc.), while subordinating others entrusted with part of health-related tasks (nurses, physiotherapists, midwives, etc.).

In the third place, the definition of a job well done refers to the internal boundaries and hierarchies within each professional group. It is on these grounds that professionals judge each other. It is not so much a question of performing the tasks requested by the public, but of doing so while respecting the conception of the professional role, and all the unwritten rules of the art. In most métiers, a professional can be disowned by colleagues for an action that satisfied the customer, and conversely, be attacked by the customer for an action deemed entirely appropriate by peers. For example, jazz musicians place great importance on preserving their freedom to perform without interference from non-musicians, especially their audiences (Becker 1985). Accordingly, they despise those who give in to "commercial" temptation and who play what "people want to hear" (p. 133). It is the maintenance of artistic integrity, of musical independence, that provides professional prestige. In the case of mathematicians, the relationship with beauty and pure pleasure makes the department of abstract mathematics the most prestigious, and the most likely to bring professional recognition. On the contrary, the department of applied mathematics, which assumes that mathematical activity is judged in function of its usefulness, constitutes a devalued field. Therefore, there are distinctive barriers within the same profession. The professional ethos embodies a standard of excellence, historically established for each métier. It orients practices, but above all, contains the elements to judge the practices of peers, "what makes us admire them, esteem them or despise them" (Zarca 2009, pp. 351–352). In this way, it is the ability to perform a job well done (which includes material and concrete dimensions, as well as social and moral ones), which grounds the hierarchy of prestige or power, to the interior of a professional group. The professional ethos goes beyond the transmission of simple rules of the art by providing practitioners with an ideal to reach and standards to judge it. This professional ideal plays out both at the individual and collective level. At the individual level, it classifies and prioritizes individuals in relation to one another. It is the work well done which determines the status of a "good" or "true" professional and organizes symbolic or practical domination relationships, ranking the best professionals over others (e.g. the primo ballerino in a corps de ballet, or a mathematician holder of a Fields Medal). At the collective level, it segments and stratifies the professional group between those whose collective practice comes the closest to the métier's ideal and those who offer a more degraded version (e.g. the corps de ballet of the Opéra de Paris versus a corps de ballet attached to a regional artistic scene, or mathematicians specializing in number theory or algebraic geometry versus statisticians). The definition of a job well done, within a professional group, can therefore be plural. While the professional ethos represents the ideal and prestigious version, it can also give rise to adaptations of the general definition, depending on profession segments. If, for an algebra specialist, statistics do not represent a way of "doing our job properly", this does not prevent the statistician from considering that they "are doing a good job" when performing a fundamentally useful task, no matter how inelegant.

Thus, the definition of a job well done and its corollary, the professional ethos, take place within a social system made up of (1) relationships between professionals and lay people, (2) relationships between professional groups and (3) relationships among members of a same professional group. The criteria for a job well done are anchored in abstract conceptions of the professional role and associate the métier with this social system, marking boundaries and hierarchies. These distinctions are not devoid of tensions; notably because the professional ethos is also an element of social identity. To state our profession does not only mean describing what we do ("I make beds, I seal teeth" in Hughes 1996, p. 95), but also involves asserting who we are based on a particular conception of labor and a "morally" hierarchical social role. Defining or redefining a job well done also means being able to say who we are. These challenges are not exempt from tensions between the professions, on the one hand, and the professionals themselves, on the other hand.

4.2. A job well done, determined by the professionals themselves

The definition of a job well done is remarkably operative and strategic for the professional group whose entire system of relationships, power and identity it organizes. But where do the establishment of the rules of the art and the constitution of the professional ethos come from? Who is to judge that a job is well done?

Traditionally, the rules of the art come from within the profession and are transmitted to new entrants via socialization processes, which lead them to internalize the conception of their role and the definition of a job well done (section 4.2.1). These rules are developed by the profession itself, in complete autonomy (section 4.2.2). They give rise to sanctions (whether regulatory or social) for those who do not respect them, and these sanctions are in the hands of the professionals themselves (section 4.2.3). These rules of the art and the definition they embrace of a job well done are developed autonomously, but not without taking into account the professions' environment. In fact, the latter address their definition of a job well done to the audiences, the state, users and other workers from whom they expect recognition (section 4.2.4).

4.2.1. *Rules of the art learned through professional socialization*

The rules of the art are learned through professional socialization. New entrants to a métier are not only trained in the techniques, but in all the ways of applying them in relation to the métier's internal judgment and valuation criteria. This professional socialization takes different forms (not mutually exclusive), depending on professional groups: academic professional training, practical professional training, professional experience with peers and passage through professional institutions.

Academic professional training involves places and times for learning both theoretical and practical knowledge, organized by (and for) the profession. Most often the trainers are professionals specializing in training. Similarly to physicians, it is during training at the academy that police officers learn a new world vision and their role. Specifically, they are led to conceive the outside world as a dangerous and violent place, a world to be distrusted and against which we must defend ourselves by mastering physical constraint techniques (Pichonnaz 2017). Training transforms young recruits into police officers, but this transformation does not end at the academy.

Practical professional training "on the job" involves peer work, real-life situations and field experience, and constitutes the extension of professional socialization. As for police officers, certain knowledge and visions transmitted at the academy are reinforced, whereas others are put into perspective in the face of professional experience, far from the métier's ideal learned at the academy (Montjardet 1996). The same is true for prison guards who only learn the rules of the art in real surveillance situations, and not in the classroom. Instead of strictly applying prison techniques (searches, incident reporting, etc.), they learn how to cooperate and negotiate with the prisoners from their fellow guards. For example, misusing incident reports is considered as an indicator of a lack of professionalism on the part of a guard, because they have failed to understand that power is co-constructed in the interaction with prisoners (Malochet 2004). The introjection of the professional ethos takes place through contact with peers and alumni.

More generally, it is the relationships with other members of the métier that enable the transmission of the rules of the art, applicable to each situation. These relationships are experienced during training, in the activity shared with peers or via different professional institutions: professional bodies, professional press, professional events and meetings, etc. This form of professional socialization is particularly relevant in métiers where the professionals practice on their own or independently (such as physicians, lawyers, notaries, artists and craftsmen). For example, for art ceramists, the socialization of new entrants occurs through professional publications, professional gatherings in the form of exhibitions, annual markets or events (international meetings, symposia and professional bodies) (Bajard 2015). It is by means of these different professional institutions that aesthetic standards (making something beautiful which is also useful) and ethical standards (valorization of humility, solidarity) are developed, making it possible to define a job well done and a good professional.

4.2.2. *Rules of the art developed by the profession itself*

The various analyses conducted on regulated professions, in other words, those having legally obtained a monopoly of the activity (physicians, lawyers, notaries, accountants, etc.), show that the rules of the art are established and controlled by the profession itself. This point assumes that professionals have structured themselves in an interest group, defending the profession's viewpoint (the conception of its role and the legitimacy of the knowledge it relies on) and obtained an exclusive authorization to practice (Larson 1977; Freidson 1986). This construction is not the result of a request from society or the state, but a group project to gain recognition (Larson 1977). Once the monopoly has been obtained, the profession is autonomous for implementing the rules of work, both from the viewpoint of tasks (division of labor, task definition and operating methods, coordination), and in terms of how to become professional (recruitment conditions, training content, career rules), or the embodiment of the professional role (relationship between practical and theoretical knowledge) (Freidson 2001). Professional bodies play a central role in the formalization and institutionalization of rules. The group's professionalism comes "from within" (Evetts 2003), in the sense that what constitutes a job well done and a good professional is determined by the profession itself. These rules are used by professionals and give rise to peer assessment. It is always a professional who judges the work of one of their colleagues, on the basis of a shared definition of professionalism.

In the case of unregulated professions (*occupations*), professional groups whose boundaries are not legally closed or regulated by a professional body, but are nonetheless recognized as legitimate to take charge of activities (plumbers, prostitutes, musicians, dancers, postmen, police officers, hairdressers, etc.), the definition of professionalism is no less autonomous. For some groups, certain dimensions (training, career rules, etc.) are organized by the organizations who recruit and hire them (police officers, teachers, hairdressers, etc.). However, this training and the assessment processes which punctuate the career are centered on profession-specific knowledge, and it is generally professionals who define the knowledge to be transmitted, who ensure its transmission and evaluate its mastery. At police academies, training is mainly provided by senior police officers who have a higher hierarchical rank than recruits and specialize in teaching. The training programs are developed by police officers, who are also senior officers. The organizational hierarchy is made up of professionals. For example, peacekeepers, brigadiers, police officers, commissioners and general police inspectors are all police officers. Hierarchical superiors shape the careers of subordinates based on the definitions of the right thing to do, which are specific to each professional group. In these cases, the professional ethos is also internal to the group. It is the result of a historical construction within the group (Zarca 2009). New entrants are made familiar with it, and it also evolves over time. Even when it is weakly formalized – in professional groups where training barely relies on abstract and theoretical approaches – the professional ethos structures professional practices. It is transmitted through socialization, during learning and then through practice peer-contact. For hairdressers, learning the métier is done through training, alternating between classes at school and apprenticeship at a hairdressing salon. The courses are taught by former hairdressers, turned into teachers. However, most of the training is provided through experience at the salon, through the contact with colleagues, including those in charge of the hair salon, who are themselves hairdressers. It is through this process that beginner hairdressers incorporate a set of formal and informal hygiene rules, which prompts them to control their self-presentation and that of their colleagues, while orienting their relationships with customers, in function of categorizations relating to the (dis)taste they inspire (Desprat 2020). Here, the professional ethos they internalize (as the definition of a job well done) is a product of the professional group, which combines socialization through professional training and the work collective.

The role of this collective is fundamental. It is in a work situation, in the interaction with other professionals, that the definition of a job well done is standardized and put into practice. Explicit and targeted advice, small implicit and general remarks, humor, mockery or congratulations are all ways of expressing the professional ethos and calling for order. Depending on the composition of this work collective and its specific conditions for professional practice, the definition of a job well done may vary. In the case of hairdressers, it will be different at luxury hairdressing salons, low-cost commercial chains, salons in rural areas, etc. However, in all cases, this definition is "in the hands" of the professionals. In that sense, professionalism is to be determined by the professionals themselves.

This professionalism is not fixed. It is also transformed following changes in context, techniques, relationships with other professional groups, demographic modifications, etc. Professionalism is a product of the professional group, which evolves with it and with its mandate.

4.2.3. *Sanctions for "poorly done" work*

While the definition of a job well done acts as a socializing norm to frame the métier's ideal, it also grounds the sanction processes against those who stray from such an ideal. In most professional services, work quality is only observed while the job is being done or ex post, once the service has been delivered. Professional services (hair-cutting, operating on a patient, defending an accused in trial, repairing a boiler, laying tiles, etc.) are singularities (Karpik 2007). These cannot be judged through the comparison between standard products or services with one another. Judging the quality of products and services is done by assessing – beforehand – the quality of those who produce or deliver them; in other words, by judging their professionalism. As a matter of fact, exhibiting the quality or title as a "professional" (being a hairdresser, surgeon, lawyer, plumber, tiler, etc.) does not suffice to determine whether the latter properly applies the rules of the art, or has internalized the professional ethos. Professionalism itself is subject to uncertainties.

To reduce these uncertainties, professionals use "personal" judgment systems (Karpik 2007). They have the particularity of relying on the persons themselves: it is the knowledge that circulates about professionals and among professionals, within networks, which makes it possible to formulate judgment, transformed into reputation, based on anecdotes, testimonies, information, rumors, etc. These interpersonal relationships bond professionals who share the same judgment criteria, the same professional ethos and who can trust one another. For example, trust built within lawyer interpersonal networks enables them to recommend colleagues to customers seeking services, while assisting customers in finding "good" lawyers (Karpik 1995). For finance professionals, the experiences of working together on certain financial operations lead them to judge on the professionalism of their colleagues. Even though these may be sometimes "adversaries" in certain operations, they will probably seek to work with them in the future, in the same camp, on other operations (Boussard et al. 2019). The different moments where professionals meet, from the most formal (such as working together) to the most informal ones (such as professional or social events), are spaces where opinions and judgment on each other's professionalism are exchanged. In the construction sector, the reputation of professionals is judged over the work performed, which then circulates in circles of professional acquaintance, facilitating the hiring of those who are best appreciated (Deffontaines 2021).

The judgment constructed within networks produces a reunion of professionals, among those who consider themselves as "good" professionals. This pairing also has the particularity of strengthening the individual reputation of each group member. This is particularly the case for artists, for example, film directors, who, by working with other professionals (direction of photography, script, editing, music, etc.) with an equivalent reputation, maintain or build their own (Menger 2002).

On the other hand, professionals who fail to be recognized as such by their peers are sent to the group's margins, where they can, by associating with others, get together as second-best professionals (e.g. specializing in particular customers or types of services). Among finance professionals, a second-best reputation is associated with a drift from the professional ethos and a specialization in smaller or less prestigious financial operations, with partners who are themselves less reputable. The sanction associated with the judgment on professionalism underlies the professional group's hierarchization, its organization into stratified networks of mutual knowledge, recurring collaborations and a division of labor in activity or clientele segments (Boussard 2015; Boussard et al. 2019).

This judgment essentially concerns the professional ethos. Judgment may fundamentally relate to mastery and compliance with ethical rules, and give rise to more severe consequences. Certain professions, in particular those which are established, regulated and have earned the State's recognition to practice an activity exclusively, have adopted a set of rules the professionals undertake to respect and which the profession's regulatory bodies (order, associations, bodies, etc.) promise to control and to sanction in case of offense. In extreme cases, the sanction implies ruling out practitioners who turn out to be recognized for not respecting professional practices. Here, the question no longer involves judging a job well done or the degrees in its execution, but rather judging "poorly done" work, to the point that the person who executes it cannot hold the professional status anymore. In the case of these regulated professions, the judgment for non-compliance with institutionalized professional rules is completely formal. These are the associations which, while being authorized to issue authorizations to practice, can also withdraw them. The various bar associations receive complaints for ethical offenses which are looked into by the president of the Bar Association. Depending on the degree of seriousness, they can sanction with "paternal admonition", in the sense of a simple remonstrance, or resort to a disciplinary procedure, whose penalty can go as far as dismissal, including warnings, reprimands or the ban on practice for a certain time (Chaserant and Harnay 2016). The disciplinary procedure involves going before an authoritative body, made up of lawyers, which investigates and judges.

In the case of unregulated professions, ethical rules are not necessarily made explicit, or may only be declarations of intent, without institutionalized procedures to ensure their compliance. For example, professionals in "financialized real estate" have adopted a code of ethics (so as not to go against the interests of their customers), but the association which brings together these professionals does not have the legal means to ban a professional from practice (Duros 2022). However, certain rules of the art can be related to ethical rules, whose non-compliance, when known to peers, is accompanied by a more informal sanction, such as marginalization. Certain practices are not allowed, even when they do not contravene explicit rules: among firefighters, the absence of solidarity with colleagues, or the inability to accept with good grace the mockery of colleagues can quickly lead to being left aside by work collectives (Pudal 2016). Among financiers, it is accepted that to win a deal, we must manipulate financial data, although there are limitations to manipulation. Those who violate certain limits and perform risky operations are quickly stigmatized and excluded from future operations: no "serious" professional will want to associate with them or recommend them (Boussard 2013). These exclusion processes play out at the level of the work collective, within teams, where professionalism "in action" is judged. In the construction métiers, it is when performing tasks that workers can access the (informal) title as a specialist, that is to say, capable of producing professional work. They are judged by other workers. On the contrary, those who cannot demonstrate they are doing a good job are confined to laborer tasks, which are arduous and less well paid. This reputation is replayed at each construction site (Deffontaines 2021). This judgment can then go beyond the framework of the work collective, circulate within networks of mutual acquaintance, and ultimately lead to a damaged professional reputation, meaning further exclusion from the professional group (absence of job offers, collaborations or recommendations). While marginalization and exclusion from the work collective are radical, being set apart from the professional group does not amount to a ban on practice or radiation. It makes the conditions of practice more difficult, leading the professionals thus marked either to leave the profession or to continue to practice it for client segments who are less careful (or wishing to find less scrupulous professionals). Professionals are then pushed to the most devalued margins of the profession.

4.2.4. *A definition of a job well done which addresses audiences*

The professional group cannot decide on its own what work it will perform and how it will do it. It must submit its claims to those the activity is aimed at (users, customers), to other workers it interacts with to carry out the tasks, and to the state which can recognize, prohibit or frame its activity. Thus, users (the *Public*), the institutional framework for work and other workers (the *Workplace*), and the *State* constitute (Abbott 1988) are the three audiences professional groups seek recognition from. Their definition of a job well done is confronted with the demands of these three audiences, which are translated as different definitions.

Users will have their own conception of the services that should be offered, the way they should be provided, the role the professional should play, etc. Professionals seek to impose their own definition, but are nonetheless caught up in the interactions with the public and the need to take into account the expectations of the recipients of their services (even if this involves reformulating them). For example, surgeons imagine the ideal patient as the one who does not ask any questions and surrenders without reluctance and with confidence to the surgeon's hands. A good surgical procedure involves having a "smooth" relationship with patients, but leaving aside the management of their emotions or fears, tensions and crises (Zolesio 2012). However, patients may be shocked by this attitude, considering it distant and contemptuous, and may make displeasure remarks. When confronted with these challenges, surgeons (and female surgeons, in particular) may be led to adapt their definition, taking more into account the requests of patients, or conversely, may justify it, by explaining that the management of patients' emotions does not fall within their métier.

Furthermore, most professional services are provided within a complex institutional framework, which engages other workers (Hughes 1996, p. 67). For example, surgeons work with physicians from other specialties, nurses, nursing assistants, stretcher bearers, hospital directors, managers, etc. Each of these professional groups has their own conception of the services offered, their own participation in them, their role, etc. In doing so, they introduce a definition of the field of action in which the activity of all is taken: illness, care, hygiene, public service, finances, research, etc. These different definitions confront one another when all these workers are required to collaborate. Each professional group constructs its definition of a job well done, within the framework of interactions with other professionals.

Finally, professional groups interact with the state, either in the form of public authorities or public administrations. The state grants them a place and recognizes the role that professional groups boast. Either by intervening to grant an exclusive right to practice the activity (lawyers, notaries, pharmacists, etc.), which it supervises through various legal or regulatory measures (prostitutes, hairdressers, osteopaths, etc.) or simply by letting things happen without supervising or taking direct action (for example, certain finance or real estate professions), the state is a central actor with whom each professional group is obliged to interact. This is obviously also the case when professional groups are housed within the state or public services (police officers, teachers, social workers, judges, etc.).

It is among these three audiences that professional groups seek to have their jurisdiction recognized (Abbott 1988). This notion of jurisdiction designates both an activity territory made up of a certain number of tasks and a right or capacity to occupy such territory. We can clearly see that even if the definition of a job well done comes from the professional group itself, it derives directly from the recognition challenges by the different audiences. Nevertheless, the professional group maintains autonomy in the definition of a job well done. Although it interacts with the audiences, its role or definition is not prescribed by them. It seeks to be recognized, to be assigned a place, a perimeter, delimited legally or symbolically, but the proposals it formulates to do so (problem definition, abstract knowledge, practical knowledge, conception of the role, professional ethos, rules of the art, etc.) are issued by it.

4.3. When the definition of a job well done escapes professionals

This model, in which the definition of professionalism is developed to the interior of the professional group, in a recognition strategy by its audiences, can be qualified (Freidson 2001) as a "*professionalism*" model. In point of fact, this model has been called into question since the 1990s. E. Freidson brings to light two other models, one he describes as the *managerialism* model and the other as the *consumerism* model. In the managerialism model, the definition of professionalism is constructed by the hierarchical line of individuals who do not have the professionalism of those they supervise and control. In the consumerism model, the definition of a job well done is given by users or customers, in a model adjusting the supply and demand. These two models attack the professionalism model because the different dimensions of "a job well done" are no longer controlled by the professionals themselves from the inside, but externally. The organizational framework and users are no longer simple audiences the groups address with a view to confirming their own conception. The balance of power is reversed. They become bodies which seek to control the work of professionals by making them endorse a conception of the services to be provided, the role to be played, the knowledge to be invoked, the ways of acting, etc., that they developed in their place, from the outside. We can complete this table with a third source defining professionalism: the state. Since the 1980s, the state has taken an increasingly active role in the regulation and organization of professional groups. In all three cases, the traditional audiences of professions have been transformed into prescribers of professionalism. The definition of a job well done no longer comes only *from within* the group (Evetts 2003). This amounts to calling into question the autonomy of professional groups, by three types of prescriptions of a job well done: by the state (section 4.3.1), management (section 4.3.2) and users (section 4.3.3).

4.3.1. *Prescription of a job well done, by the state*

In the professionalism model presented so far, the state is seen as a third-party involved in the development of professions. It is a "secondary", "passive", "inconsistent" actor (Le Bianic and Vion 2008, p. 257) who limits itself to recognizing the efforts made by professional groups to acquire the attributes of regulated professions and have them recognized in the form of group regulation autonomy. The state is a legal partner that grants rights to competing groups. In this way, the state regulates the existence of professional groups. On the other hand, it does not intervene in their regulation and as we have seen, the definition of professionalism is therefore autonomous. State control is not exerted on professionalism itself, once a jurisdiction or a right to exist has been granted in its name.

However, if we take a closer look (Bajard et al. 2018), we may notice that the state can actually intervene in the professionalism of groups. Through the services they intend to provide to the population, professional groups can indeed give rise to public action. For example, professions intending to cure, including alternative approaches to institutionalized medicine (homeopaths, hypnotherapists, osteopaths, magnetizers, sophrologists, etc.), touch on public health issues. Public health policies can intervene proactively (rather than reactively) to encourage the existence of a profession or to require its regulation. Another example of the role of the state in the transformation of group professionalism can be found in the case of occupational medicine. When it comes to occupational health in France, it is the new public policy orientations involving multidisciplinary care that have redefined the role of occupational physicians since 2002. This was done by creating new nursing specializations (in occupational health), expanding the role of assistants in occupational health centers and giving a legitimate place to psychologists and ergonomists in preventive actions (Barlet 2019). The appearance of these new actors, or the modification of their role and their legitimacy, through state intervention, has an impact on the activity of occupational physicians. They have been led to redefine themselves as "conductors", having to coordinate other professionals. We see that this public policy does not merely listen to professional groups before granting them license and a mandate. It redefines their license and charges them with a mandate. In the same way, the public policy implemented by the French Ministry of Justice through the law known as J21, gave a mandatory role to family mediators for conflict resolution between ex-couples (in certain, specific cases). This obligation is contradictory to the definition of work specific to mediators, which requires people attempting mediation to be willing to do so. By giving a license to family mediators, and also a new activity territory, the state also intervened in their mandate. Mediators have thus redefined their activity as (also) comprising convincing ex-couples of the benefit of communicating to solve their conflicts. This new mode of intervention transforms the group's professionalism (Boussard 2021). We could mention many other examples, in several public policy fields (health, cultural, social, educational, etc.), all of which show that the state has a role as a prescriber of a job well done for some professional groups.

This prescription, which tends to control group professionalism, involves rather discreet measures, which have an indirect impact. State legislative action toward professional groups involves varied and extensive modes of action, which interweave several regulatory levels: rules for awarding public subsidies, social security code, tax code, regulations on products and services, etc. State action toward the group involves far more than the granting of a "license" (Hughes) or "jurisdiction" (Abbott). The legislative activity in question takes place at levels below that of the "super-rules" (Paradeise 1984), organizing the closing of labor markets. However, these sub-level rules have a determining impact on the definition of professionalism. For example, it is through the financing of cultural action that the state contributes to the legitimate definition of a way of being a stage director (Glas 2018). The state implements a classification work of the different structures in order to allocate subsidies. It draws a distinction between drama centers and permanent troupes, led by group leaders (most often stage directors) and cultural centers (often led by cultural facilitators). The state also establishes a hierarchization of public stages (national theaters, drama centers and companies), which structures the careers of stage directors, turning the management of a public theater into the ultimate consecration. By means of infra-level rules, the state constructs the principles of the professional ethos of the group.

Furthermore, since the neo-liberal turn of the 1980s, the state has changed its approach toward activity monopolies attributed to certain professional groups. The neoliberal economic trend treats professions as powerful mechanisms for restricting competition (Dingwal 2008). The division of labor, bans on advertising and monopolies are considered contrary to the principles of free competition, in itself perceived as a guarantee of greater market efficiency, and of goods and services obtained at the best possible quality/price ratio. In a seminal article from 1945, the economists Friedman and Kuznets (1945) proposed removing monopolies and subjecting professions, like all other types of cartel, to anti-trust laws. From this perspective, professional groups having obtained certain jurisdiction are considered as an obstacle to competition. In the United States and Great Britain, this economic perspective led to a vast deregulation movement, initiated in the early 1980s. On the European continent, this is translated as a decline in the regulatory role of Nation States in favor of the transnational level, represented by the European Commission. The harmonization of university degrees and diploma levels at the European level creates a standardization of entry conditions in certain regulated professions to the detriment of former national legislation, as, for example, with the creation of a new market for legal services (Boigeol and Dezalay 1997). In France, after having accompanied, or even encouraged in the closing of labor markets (Dubar et al. 2015), by granting jurisdictions to self-regulated professional groups from the end of the 1980s, the state undertook to liberalize the markets for goods and services, and to put an end to certain licenses, granted for long. At the end of the 1980s, this initially resulted in the liberalization of financial services and the dissolution by the state of the stockbrokers company, a professional group who had been created by means of a decree in 1801 (Godechot 2001; Riva and Lagneau-Ymonet 2012). Even more pronounced, in the mid-2000s, regulated professions were accused of maintaining unjustified monopolies from the viewpoint of the service provided to consumers and inequalities of access to the market of the employment (Report of the European Commission on Competition in Professional Services 2004–2005). In France, this culminated in the bill "for growth, activity and equal economic opportunities" (2015), which reviewed the licenses granted to certain professional groups and prompted them to redefine their mandate: solicitors, notaries, lawyers, pharmacists, taxis, etc., were affected. For the state, it was no longer a question of organizing the self-regulation of professional groups, but rather of participating in their regulation, by replacing professional rules with "market rules" (more precisely, by legislating to create specific competition conditions).

4.3.2. *Prescription of a job well done, by management*

The neoliberal trend does not only result in an intervention on the licenses granted to professional groups and on the transformation of the conditions of entry into them. It also introduces mechanisms that will make it possible to imitate market conditions where there are none in order to create a certain form of competition. In other words, the idea is to set up schemes for comparing the products and services provided, in terms of quality, price and costs, and thus improve the service provided to users and reduce costs. These practices are developed within organizations hiring professionals to bring into competition the different services and services provided. They rely on indicators (productivity, quality, cost, etc.) as a means to assess the work carried out by professionals. These indicators are accompanied by the setting of targets to be achieved. The comparison involves the standardization of services or modes of action to be implemented (skills, good practices, procedures, etc.), which leads to prescribing the means to be used by the professionals. These organizational schemes are part of a managerial logic, initially developed in private companies, which sought, from the outset, to put employees under the control of management (Boussard 2008). Management systems have the particularity of reversing the traditional work order, in which workers, experts in their métier, organize the progress of their task, shielded from the gaze of their contractor. In this traditional organization, workers maintain autonomy over the way they perform their work, while the contractor receives a service or finished product. Managerial systems break down this separation by enabling the contractor to precisely prescribe the task progress of the workers they hire. Knowledge about work shifts to the side of employers and management. As Freidson (2001) points out, it is the managerial logic, rather than professionals, that organizes the control of work: division of labor, definition of tasks and missions, coordination, recruitment conditions, career rules, training content, the relationship between practical and theoretical knowledge no longer being decided autonomously by professional groups, but prescribed by the hierarchy (specialists in work control), the managers. Used massively since the beginning of the 20th century in the industry, these schemes would keep the various professions inherited from the corporations which populated the industrial workshops under control. After the end of the 20th century (1980s), these systems massively invested in service activities in private companies (Power 1999; Bruno and Didier 2013).

The bureaucratic knowledge that such schemes implemented was presented as general knowledge, overlooking and delegitimizing the particular knowledge of each professional group (Freidson 2001). This general knowledge was also transnational in nature, applicable to various local contexts. It was at the origin of a whole range of standardized tools which themselves aimed to standardize professional work: approaches to quantifying activity in the form of indicators and dashboards, activity standardization and certification procedures such as ISO procedures, activity control procedures via IT and digital tools (ERP software, digital applications, etc.). For example, the activity of truck drivers was profoundly affected after the arrival of route optimization software and geolocation applications. Route optimization software made them lose autonomy over the organization of their itinerary, while GPS applications made them lose control over its implementation (Maugeri 2003). In eateries belonging to chain restaurants, the various schemes (performance control, quality, hygiene standards, presentation and customer relations, etc.) frame and constrain the definition of a job well done by each of the cooks and waiters (Dujarier 2006).

For many professional groups in these companies (accountants, logisticians, salespeople, etc.), the development of managerialism is accompanied by a transformation of the definition of a job well done. Indeed, the approaches used require professionals to control the means and ends of their activity. Not only do the schemes set targets at work (improving quality, increasing customer satisfaction, reducing costs, etc.) but they also supervise the means to be used by defining the procedures to be followed. The schemes also have the purpose of controlling target achievement and the monitoring of procedures, which shifts control over work from within the professional group to its exterior, and more precisely, to the hierarchy ("from above"). Evetts defines professionalism coming from above as organizational professionalism, in the sense that it is defined not by the professional group, but by the organization which hires and controls it (Evetts 2003). In the organizational professionalism model, the profession loses its capacity for self-determination over the rules of the art.

This managerial mode of work control in for-profit organizations has been extended, in its contemporary forms, to public administrations, in a context of budgetary restrictions and a neoliberal perspective. Baptized as the New Public Management (NPM), this doctrine or "doctrinal puzzle" (Bezes 2009) presents a solution to what appears to be a central problem in the development of an administrative state (public deficits, internal dysfunctions, lack of efficiency, increasing costs, etc.). As they deal with health, education, justice, police, social affairs or culture in the name of the state, many administrations also hire professionals: physicians and nurses, teachers, magistrates and court clerks, police officers, educators and social workers, musicians, dancers and actresses, etc. As it reuses standardized management approaches, New Public Management places these professional groups in an *accountability* scenario, transforming professionalism into organizational professionalism. For example, the implementation of quality procedures in hospitals, introducing multiple measurement indicators, modifies the professional practices of physicians, as in the attention paid to keeping patient records (Bertillot 2016). Even if they are negotiated by the managers in charge of quality control with the physicians, these indicators do not correspond to their professional definition of a job well done. For physicians subject to quality control, the definition of good medical work is shifting toward a new standard, stemming from the organization, which points to proper administration while keeping patient files. As a matter of fact, in the case of a surgeon, this can be contradictory with the widely shared definition of a good surgeon as somebody who is good at operating, gives suitable indications and has developed the technical gesture (Zolesio 2012).

In many public services hiring professionals, the hierarchical line is itself made up of professionals. Public highway police officers are led by other police officers, brigadiers, officers and commissioners. Magistrates are also organized and supervised by the president of a tribunal or court, themselves a magistrate. We could multiply the examples. The professionals in charge of leading their peers are encouraged to use managerial schemes to carry out their mission (supervision, organization and control). Professional knowledge is replaced by the managerial logic, which introduces new standards for judging a job well done. We can see this in the police commissioners who use performance indicators developed by the ministry to orient the activity of brigades, by setting performance targets (for example, number of arrests or case resolutions). Despite this, such targets do not correspond to the idea that public highway police officers have of their métier, because they lead them to stop cars just "to make numbers" and not out of a concern for public safety (Boussard 2010; Douillet et al. 2014). While being operated by professionals, professionalism embracing this managerial logic remains fundamentally an external construction to the professional group.

Thus, the managerialism model, whether taking place in for-profit organizations, public administrations or non-profit organizations (Hély 2009), produces an "injunction to professionalism" (Boussard 2010), by proposing or imposing new definitions of a job well done, from the outside. This injunction to professionalism is made to professionals through the audiences constituted by other workers in charge of evaluating their activity (the *workplace*). This concerns the hierarchical line, in the first place, but they can also be other workers, who work upstream or downstream the chain of command. These peers can take advantage of the managerial schemes put in place by the organization to implement new ways of doing work or controlling it: physicians in relation to nursing assistants, salespeople in relation to production or logistics services, magistrates toward educators, police officers toward magistrates, etc. As in the state, the other workers involved are far more than simple audiences. They are actors who, thanks to managerial schemes, have the power to redefine the role, knowledge and more generally, the ethos of the professional groups they interact with.

4.3.3. *Prescription of a job well done, by users*

The users, who are one of the other audiences, are caught in the same dynamic. They are the ones who begin to control work (Freidson 2001). Still, from a neoliberal perspective making the market the Alpha and Omega of all action, users (seen as consumers of services and benefits) are judged as being in the best position to know what suits them. They are considered as individuals having the knowledge and abilities that enable them to make choices in their best interests. For this, they do not need to rely on the knowledge of specialists, or professionals. Their own assessment of situations, based on their ordinary knowledge, is the ultimate reference for all choices. Under these conditions, the specialized knowledge of professionals must be at the service of consumers who are able to judge as to its relevance.

In this *consumerism* model, the market, embodied by the judgment of users and customers, becomes the logic capable of defining a job well done. This market logic takes the form of two distinct logics, which can overlap: a strictly market logic in one case, in which the performance of professionals is judged by a price. It is then the economic dimensions of the service which prevail in matters of judgment, putting aside all the procedural dimensions, related to the respect for the rules of the art. Since work control is only evaluated on the economic grounds of the product or service, it disintegrates any notion of a job well done, other than the perspective proposed by the productivist approach. The second logic is a profane one: it introduces non-professional, distant criteria into the judgment of the services provided, independently from the subtleties of the work carried out. In the name of their own choice criteria, users may be led to request services, which do not correspond to what professionals consider to be good work: the patient who dictates to a physician the medications they must prescribe, the customer who reprimands their lawyer on the legal arguments they should use, the public who require musicians to play a certain program, etc. In a nutshell, satisfying the customer can amount to ignoring all the rules of the art and the ethos that traditionally builds up a profession.

However, for commercial or profane logic to have an effect on professions, they must be able to rely on the schemes that make them possible and visible. We have already seen that when it comes to professional services, there is radical uncertainty about quality. In point of fact, for market logic to work, products with similar qualities must be able to be compared on the basis of a price. For profane logic to also work, users must have all the information to compare services with one another. Actually, for the market of singularities constituted by professional services to exist, it has to rely on comparison mechanisms, which enable users to choose the "best" service, as per their judgment criteria. While in the professionalism logic these personal judgment schemes produce concrete knowledge regarding the professionals who offer services, in the market's logic, it is the impersonal judgment schemes (Karpik 2007) that enable the comparison of professionals and services. These impersonal schemes come in the form of ratings, rankings, labels, prizes (rewards), etc. They may be derived from ratings made by users themselves, which are then aggregated (service ratings on digital platforms), or from rankings made by specialized intermediaries (media, guides, etc.). In both cases, the indicators produced to judge the services introduce external definitions of a job well done, distant from those of professionals, leading to a transformation of their practices in order to succeed in producing services consistent with the criteria measured by said indicators. Just to mention an example: following the ranking of law universities in the United States, according to standardized criteria, including the existence of data on student employment after graduation, the staff from student placement services began to spend more time conducting surveys on the future of their students, than on supporting them in their job search (Espeland and Sauders 2007).

It is important to stress that the judgment criteria used are rarely those of the users themselves, but are criteria designed by intermediaries, to inform on the users' judgment. Paradoxically, the market logic assumes the existence of experts in impersonal judgment schemes, who intervene between users and professionals (Hatchuel 1995, pp. 205–225; Bessy and Chauvin 2013): intermediaries who publish rankings, guide editors (Parker wine guide, Michelin guide), designers of rating algorithms for digital platforms, etc. It is rather the latter, than the "market" itself, which prescribe the definitions of a job well done through criteria they put in place to inform on the users' judgment.

Another way for users to prescribe a job well done is to go deeper into the activity of professionals by offering the professionals the knowledge they should master in the context of their activity. This involves developing a form of expertise which competes with that of professionals, calling them into question and the power they exert over users in virtue of their expertise. It is not so much the dispersed users facing professionals who intervene on this point, but rather the user associations or "expert" users who speak on their behalf. In the case of neonatal resuscitation, it is with a parents' association that physicians are required to discuss the criteria followed for deciding on not resuscitating a premature child (Paillet 2007). The "profane" expertise they construct is based on different non-exclusive forms. It can involve re-appropriating and rereading scientific or technical work, developing controversies (Paicheler 2010), entering into confrontation with the "orthodox" knowledge of professionals. It can also be expertise built on the basis of user experience, which claims to be more legitimate than professional knowledge and is used for defining the way in which professional work should be done. The health, social or educational professions are particularly worried by the development of this profane expertise (Carricaburu and Ménoret 2004).

This situation (prescribing a job well done to professionals) is reinforced by the use of the Internet and social networks which makes this "profane" expertise available to users who can then use it in the context of their interactions with professionals, thereby calling their authority into question and putting pressure on them to conform practices to their own representations of what they should be like. The most classic example is that of the patient who relies on the information they read on the Internet to suggest both a diagnosis and a treatment to their treating physician.

4.4. Struggles around the definition of a job well done

The intervention of audiences in the definition of a job well done by professionals leads to stopping seeing the relationship between the latter and professional groups as a simple support for recognition. Professional groups are part of configurations of multiple actors who are not there only to recognize the standards developed internally by the group, but who actively participate (to varying degrees), in the normative construction of the definition of a job well done (Boussard 2010). At the intersection of these different interrelations, professional groups find it increasingly difficult to maintain an autonomous and shared definition of a job well done. As a matter of fact, its own ability to define a job well done is a fundamental dimension of the autonomy of a professional group. Calling autonomy into question results in struggles, led by professional groups, to control and assert their own definition.

Struggles for professional autonomy and struggles to maintain control over the definition of a job well done depend on these power relations, which are organized following two models. The first corresponds to institutional struggles, which involve symbolic battles, interest defense groups and are based on the mobilization of scientific and moral discourses, aiming to establish the group's viewpoint on the definition of a job well done. The second model corresponds to pragmatic struggles, which take place on the very terrain of activity and are barely anchored in the construction of collectives and often operate clandestinely.

4.4.1. *Institutional struggles*

In this first model, the defense of professional autonomy implies a defense of institutional autonomy. The professional group seeks to constitute itself as an institution to produce a discourse about itself and its social role, which becomes a reference for the other actors in the configuration. In this view, the struggle is a political one, which turns the professional group into an interest group (Offerlé 1998), seeking to build its viewpoint, have it recognized and defended. This requires the group to have spokespersons capable of constructing a collective position and identifying the interests to be defended. Associations, orders, professional bodies, occupational unions and networks are all more or less legally formalized forms, which provide the group with the capacity to develop and assert its own role, its own definition of a job well done, within the professional configuration. There we find notaries, magistrates, physicians, teachers, police officers, nurses, specialized educators, social workers, accountants, company treasurers, osteopaths, pharmacists, etc. By structuring an official definition, their spokespersons help to build a collective position shared by group members.

This institutional work of the professional group involves both symbolic dimensions, and practical or legal ones. It is both a matter of recognizing the existence of the group and the legitimacy of its license and its mandate, and equipping it with (or defending) the concrete guarantees of this recognition (monopoly or right to operate, financing, entry methods into the group, career rules, remuneration, etc.). This institutional work is particularly visible when professional autonomy is attacked by a new regulatory context, or the transformation of public action. This is the case for notaries, whose monopoly on the drafting and authentication of legal documents has been regularly attacked since the 1960s, and a series of European pressures encouraging the liberalization of legal services in Europe, which has been severely contested since the 2000s. However, the French

Law No. 2015-990 from August 6, 2015 "for growth, activity and equal economic opportunities" (known as the "Macron law") introduced changes in the access to the profession and its practice, without, however, calling its monopoly into question. The Higher Council of Notaries, a professional organization of notaries responsible for representing them before the public authorities, constructed an argument based on the defense of legal security and a continental legal order threatened by the legal order from English-speaking countries, which was supported both by the French Ministry of Justice and groups of academics and legal professionals (Delmas 2019). This mobilization of the professional group, via its representatives, is an example of institutional work carried out by a professional group to fight against the challenges to its autonomy.

In these struggles for professional autonomy, the group can rely on former alliances with other actors in the configuration, within what Abbott (2003, pp. 29–50) called "linked ecologies". This is evident in the example of notaries who rely on other professional legal groups to defend their positions. The relationships that the group develops with other actors, and in particular other professional groups, can allow it to adopt observation positions (Hénaut 2011) regarding the way in which the latter defend their own professional autonomy, which can act as a source of inspiration and imitation.

The group can also go through other relational processes which are its "avatars" (Abbott 2003, pp. 29–50), that is to say, former group members who, during their career, joined other actors in the configuration: ex-professionals who became members of a state administration, in charge of political or governmental responsibilities, members of the hierarchical line or of a user association, etc. For example, stage directors (Glas 2018) or police officers (Pichonnaz 2017) are invited to accept positions within the administration and ministries from where to act over their own professional groups of origin.

Confronted with attempts to intervene on the definition of a job well done proposed by managerialism, groups can seek to integrate their own criteria into assessment methods and thus re-appropriate the indicators, or their interpretation. In courts subject to New Public Management – which organizes managerial control, essentially of case processing times and court organization methods – many judges consider that an evaluation of their activity is necessary and that respecting user deadlines is of the utmost importance. Nevertheless, they re-appropriate the injunctions and methods given to them by adapting them to their own definition of a job well done (Vigour 2008). From this viewpoint, the most institutionalized groups are not without resources to reinvest managerial logics with their own logics. However, these reinvestments and diversions of managerial logic are not exempt of effects on the professional group: the use of indicators, standardized procedures or impersonal judgment systems leads to the formalization of new definitions of a job well done which can create tensions, cleavages and segmentations within the group. Among judges, for example, the new criteria for judging a job well done, focusing on shortening deadlines and adapting criminal responses in order to reduce pending cases, strengthen the prerogatives of public prosecutors (*magistrat.es du parquet*) compared to those from the *magistrat.es du siège* (Vigour 2008; Milburn et al. 2010). Generally speaking, definitional struggles do not leave professional groups unscathed: they lead them to recompose or transform themselves, in relation to the balance of power within the configuration where they are involved, and the power dynamics to the interior of the group itself. The spokespersons' constitution of an institutional definition of a job well done, with a view to building and defending the group's interest, can come into contradiction and conflict with definitions of a job well done shared by the group members, placed at other levels (in contact with users, related professional groups, hierarchical line, etc.). The plural definitions of work within professional groups, at the basis of segmentation, will relate to external prescriptions, and through recomposition, create new segmentations or accentuate existing ones.

4.4.2. *Pragmatic struggles*

Opposite the institutional struggles model is that of pragmatic struggles. This concerns professional groups in a subordinate position within a work organization or a labor market, lacking collective structures specifically dedicated to the defense of the professional ideal. The spectrum of activities targeted is wide: professional workers, technicians, handlers, delivery men, vehicle drivers, postmen, caregivers, waiters and waitresses, beauticians, hairdressers, salesmen, security guards, carers, childminders, etc.

For each of these groups, the definition of a job well done befalls onto employers (including if they are themselves professionals such as beauticians, hairdressers, etc.) or customers. This definition is translated into different managerial schemes for standardization, control and assessment, which tend to confine the activity within precise frameworks. They prevent the implementation of a definition of a job well done developed in accordance with the activity itself, which gives sense to it. This "impeded work" (Clot 2010) gives rise to the accomplishment of tasks which are seen as "dirty work", not in the sense of degrading, downgraded tasks, but as degraded tasks. Professionals have the feeling of doing a poorly done job, a job that contravenes the rules of the art. This is, for example, the case of postmen, whose activity is increasingly constrained by time, following the calibration of itineraries by software, and thus feel obliged to give up contact with users (Demazière and Mercier 2013). We can also mention the case of hairdressers who in certain salons must systematically offer additional treatments (on site or to be purchased), which thwarts the disinterested advice relationship they seek to establish with their customers (Desprat 2017).

The particularity of these definitions is that they are locally constructed and shared, within the framework of a work collective. They are generally little or not supported by an argument beyond the collective's framework While employee mobilizations exist, they entice local labor conflicts. Even if some of these mobilizations can have a national resonance (strike of caregivers, bicycle delivery men or Uber vehicle drivers), they are not conducive to a national and powerful representation of the group, capable of asserting their definition of a job well done. We can also note that for these groups of workers, mobilization often concerns remuneration and working conditions issues, rather than professional autonomy challenges.

In this situation, where there is little or no room for recognition of the professional group and a shared definition of a job well done, the struggles to defend professionalism occur in the acts of work themselves. It is not so much a question producing a discourse on what labor should be, but of trying to do the work as it should be done. These attempts to maintain work within the framework of the rules of the art are made against the requests of hierarchy or customers. They give rise to tinkering, leeway gained here and there. They are the subject of small victories which are more akin to autonomy at work (de Terssac 1992) than to professional autonomy. The above-mentioned postmen or hairdressers manage to resist the injunctions: some try to save time during their itinerary to spend more time with certain users, whereas others will only offer special treatment to certain customers, so as not to pollute the relationship with loyal customers. However, in a situation where these arrangements amount to cheating the system, they are risky and need to be carried out clandestinely (Borzeix and Linhart 1988; Espeland and Sauder 2007). The latter leads to the isolation of each worker, who will fight in their corner to try to maintain the rules of the art and the meaning in their work. Even if clandestineness can lead to collectives, these are strictly localized, part of sharing the same activity. By nature, they cannot invite the publicization of the work done and therefore to the collective and official defense of their definition of a job well done.

These pragmatic struggles make work well done exist on a daily basis, against the various attacks it suffers, but this existence is circumscribed and clandestine, in permanent tension with the exogenous definitions imposed on professionals. Note that this defense of work well done in actions also exists in the institutional struggles model. However, it is accompanied by a discursive, constructed and institutionalized defense of it. Finally, notice that the two struggling models surrounding a job well done are the two poles of a continuum, and as such, do not constitute exclusive models. Many groups can borrow traits from one or the other, and present a singular mixture, depending on the characteristics of their configuration and their political, economic, technical and demographic situation. They can also slide along the continuum over time, some moving from an institutional struggle to a more pragmatic one (as postmen do, Demazière and Mercier 2013), others managing to transform pragmatism into institutionalized discourse (restorers of works of art, Hénault 2011).

4.5. Conclusion

The definition of a job well done is at the heart of the very existence of professional groups and their claim to professionalism. It is through it that the group can exist: it brings together the workers who share it, distinguishing them from those to whom this work is addressed and from those whose work can be defined differently. Even more fundamentally, the definition of a job well done organizes relationships within the professional group by creating polarizations, internal segmentations and hierarchies, which have repercussions through a process of work division and delegation, as well as on the hierarchies between professional groups operating within the same activity space. The definitions of a job well done feed struggles between professional groups which account for their competition to occupy a place and a role in the division of labor and the distribution of activity territories.

Within the same professional group, the definition of work is underpinned by an ideal, but this gives rise to a plurality of interpretations, depending on the position of professionals in the internal segmentation. Tensions therefore exist around this definition. However, these tensions are the manifestation of internal dynamics within the professional group. They correspond to challenges regarding boundaries and symbolic hierarchies within the group, and give rise to visible struggles, both in the work collectives and at the more institutional level of professional socialization bodies.

Another type of conflict around the definition of a job well done brings together professional groups and their audiences, which, over the past 30 years, have shifted from being instances of recognition and legitimization of the claims to professionalism from professional groups to prescribing instances. In these cases, the definition of a job well done no longer comes from within the professional group (developed, transmitted, controlled by professionals), but is external. The state, management, users and customers become those who, from outside the group, impose a definition of a job well done and often verify its application. Confronted with these prescriptions, professional groups have to fight to maintain control over their definition of a job well done, and thus maintain their autonomy. The form of these struggles and the modalities of preserved autonomy are organized according to two polarities: institutional struggles for professional groups able to make their arguments on the definition of work heard and recognized through spokespersons, and pragmatic struggles for rather subordinate professional groups in the division of labor, at the more concrete level of work in action. However, the two forms of struggle are not mutually exclusive. These struggles are not unrelated to the internal tensions that run through professional groups regarding the definition of a job well done. They either rely on them or feed them. In that sense, they participate in recomposing professional groups and their internal segmentation. They have consequences on relations between professional groups and the division of labor.

It is nonetheless clear that groups which fail to mobilize collectively to assert and defend their professionalism are the most vulnerable to attacks coming either from competing groups or from prescribers of professionalism. The mobilization of professional groups is essential to their existence and its continuation, and all organizations which destroy the possibility of developing work collectives and the representation of interests accentuate such vulnerability. Despite this, mobilization, even when conducted in the sense of the general interest, is always likely to be qualified as "corporatism", in the sense of an action aimed at defending the interests of the constituted group. The mobilizations of those who seek to prescribe their professionalism, in the name of the general interest (whether led by the state, management or users), are more rarely perceived as the defendants of particular interests. It is undoubtedly one of the effects of the (neo)liberal discourse to have introduced this interpretation. The sociology of professional groups – while continuing to show the processes through which the positions of professional groups are constructed – also has the task of deconstructing the arguments of the prescribers of professionalism and analyzing how the latter can also be the fruit of professional groups and related professional ecologies (Abbott 2003, pp. 29–50). The state, management and user associations are also populated by professional groups.

4.6. References

Abbott, A. (1988). *The System of Professions: An Essay on the Division of Expert Labor*. University of Chicago Press, Chicago.

Abbott, A. (2003). Écologies liées : à propos du système des professions. In *Les professions et leurs sociologues. Modèles théoriques, catégorisations, évolutions*, Menger, P.-M. (ed.). Editions de la MSH, Paris.

Arborio, A. (2009). Les aides-soignantes à l'hôpital. Délégation et professionnalisation autour du sale boulot. In *Sociologie des groupes professionnels. Acquis récents et nouveaux défis*, Demazière, D. and Gadéa, C. (eds). La Découverte, Paris.

Bajard, F. (2015). L'invention de la céramique d'art. Contribution à la sociologie de la construction des professionnels. *Sociologie du travail*, 57(3), 299–321.

Bajard, F., Crunel, B., Frau, C., Nicolas, F., Parent, F. (eds) (2018). *Professionnalisations et Etat. Une sociologie politique des groupes professionnels*. Presses du Septentrion, Lille.

Barlet, B. (2019). *La santé au travail en danger. Dépolitisation et gestionnarisation de la prévention des risques professionnels*. Octarès, Toulouse.

Becker, H.S., Hughes, E.C., Geer, B., Strauss, A.L. (1961a). *The Making of a Physician in Boys in White: Student Culture in Medical School*. University of Chicago Press, Toronto.

Becker, H.S., Hughes, E.C., Geer, B., Strauss, A.L. (1961b). *Boys in White: Student Culture in Medical School*. University of Chicago Press, Toronto.

Becker, H.S. (1985 [1963]). *Outsiders. Études de sociologie de la déviance*. Métailié, Paris.

Bernardi, P., Maitte, C., Rivière, F. (eds) (2020). *Dans les règles du métier. Les acteurs des normes professionnelles au Moyen Âge et à l'époque moderne*. New Digital Frontiers, Palermo.

Bertillot, H. (2016). Des indicateurs pour gouverner la qualité hospitalière. Sociogenèse d'une rationalisation en douceur. *Sociologie du travail*, 58(3), 227–252.

Bessy, C. and Chauvin, P.M. (2013). The power of market intermediaries: From information to valuation processes. *Valuation Studies*, 1(1), 83–117.

Bezes, P. (2009). *Réinventer l'Etat. Les réformes de l'administration française (1962-2008)*. PUF, Paris.

Boigeol, A. and Dezalay, Y. (1997). De l'agent d'affaires au Barreau : les conseils juridiques et la construction d'un espace professionnel. *Genèses*, 49–68.

Borzeix, A. and Linhart, D. (1988). La participation : un clair-obscur. *Sociologie du travail*, 30(1), 37–53.

Boussard, V. (2008). *Sociologie de la gestion. Les faiseurs de performance*. Belin, Paris.

Boussard, V. (2010). Les policiers de voie publique et l'injonction à faire du chiffre. In *L'injonction au professionnalisme*, Boussard, V., Demazière, D., Milburn, P. (eds). Presses Universitaires de Rennes, Paris.

Boussard, V. (2013). Qui crée la création de valeur ? *La nouvelle revue du travail* [Online]. Available at: http://nrt.revues.org/1020.

Boussard, V. (2015). De l'espace hybride au monde social : la coordination dans le marché en abyme des transactions d'entreprise. *L'année sociologique*, 2, 357–390.

Boussard, V. (2018). Professional closure regimes in the global age: The boundary work of professional services specializing in mergers and acquisitions. *Journal of Professions and Organization*, 5(3), 279–296.

Boussard, V. (ed.) (2021). L'évaluation de la tentative de médiation familiale préalable obligatoire (TMFPO). Quand médier n'est pas remédier. Report, Mission Droit et Justice, GIP Justice, Paris [Online]. Available at: http://www.gip-recherche-justice.fr/wp-content/uploads/2021/01/18-21-Rapport-final-TMFPO-janvier-2021-annexes-et-resume-fr-optimis%C3%A9-1.pdf.

Boussard, V., Godechot, O., Woloszko, N. (2019). How to make a deal: The role of rankings and personal ties in creating trust in the mergers and acquisitions market. *Socio-Economic Review*, 17(2), 311–336.

Bruno, I. and Didier, E. (2013). *Benchmarking. L'État sous pression statistique*. La Découverte, Paris.

Carricaburu, D. and Ménoret, M. (2004). *Sociologie de la santé. Institutions, professions et maladies*. Armand Colin, Paris.

Chadouin, O. (2021). *Sociologie de l'architecture et des architectes*. Editions Parenthèses, Marseille.

Chaserant, C. and Harnay, S. (2016). La déontologie professionnelle en pratique. Enquête sur l'activité disciplinaire de la profession d'avocat. *Revue Française de Socio-Économie*, 16, 119–139.

Clot, Y. (2010). *Le travail à coeur. Pour en finir avec les risques psychosociaux*. La Découverte, Paris.

Deffontaines, P. (2021). Au-delà de l'informel : réputation et normes sur les chantiers de construction en Ukraine. *Sociologie du travail*, 63(1).

Delmas, C. (2019). *Les notaires en France. Des officiers de l'authentique entre héritage et modernité*. Presses Universitaires de Rennes, Rennes.

Demazière, D. and Gadéa, C. (eds) (2009). *Sociologie des groupes professionnels. Acquis récents et nouveaux défis*. La Découverte, Paris.

Demazière, D. and Mercier, D. (2013). La tournée des facteurs : normes gestionnaires, régulation collective et stratégies d'activité. *Sociologie du travail*, 45(2), 237–258.

Desprat, D. (2017). "Qu'est-ce qu'on vous fait aujourd'hui ?" Un ethos professionnel des coiffeurs. Entre travail émotionnel, relation de service et dispositions genrées et de classe. Thesis, Université Paris Nanterre.

Dingwall, R. (2008). *Essays on Professions*. MPG Books, Bodwin.

Douillet, A.-C., de Maillard, J., Zagrodzki, M. (2014). Une centralisation renforcée par le chiffre ? Les effets contradictoires des indicateurs chiffrés dans la police nationale en France. *Politiques et management public*, 31(4), 421–442.

Dubar, C., Tripier, P., Boussard, V. (2015). *Sociologie des professions*. Armand Colin, Paris.

Dujarier, M.-A. (2006). *L'idéal au travail*. PUF, Paris.

Duros, M. (2022). L'édifice de la valeur. Sociologie de la financiarisation de l'immobilier en France (fin des années 1980-2019). Thesis, EHESS, France.

Espeland, W. and Sauder, M. (2007). Rankings and reactivity: How public measures recreate social worlds. *American Journal of Sociology*, 113(1), 1–40.

Evetts, J. (2003). The sociological analysis of professionalism. Occupational change in the modern world. *International Sociology*, 18(2), 395–415.

Friedman, M. and Kuznets, S. (1945). *Income from Independent Professional Practice*. National Bureau of Economic Research, New York.

Freidson, E. (1986). *Professional Powers: A Study of Institutionalization of Formal Knowledge*. University of Chicago Press, Chicago.

Freidson, E. (2001). *Professionalism, the Third Logic*. Polity Press, Cambridge.

Glas, M. (2018). Se professionnaliser avec ou contre l'État : le metteur en scène entre revendications sectorielles et institutionnalisation du théâtre public. In *Professionnalisations et Etat. Une sociologie politique des groupes professionnels*, Bajard, F., Crunel, B., Frau, C., Nicolas, F., Parent, F. (eds). Presses du Septentrion, Lille.

Godechot, O. (2001). *Les traders. Essai de sociologie des marchés financiers*. La Découverte, Paris.

Hatchuel, A. (1995). Les marchés à prescripteurs. Crise de l'échange et genèse sociale. In *L'Inscription sociale du marché*, Jacob, A. and Vérin, H. (eds). L'Harmattan, Paris.

Hély, M. (2009). *Les métamorphoses du monde associatif*. Presses Universitaires de France, Paris.

Hénaut, L. (2011). Capacités d'observation et dynamique des groupes professionnels. *Revue française de sociologie*, 52(1), 71–101.

Hughes, E.C. (1958). *Men and Their Work*. Free Press, Glencoe, IL.

Hughes, E.C. (1996). *Le regard sociologique. Essais sociologiques*. Éditions de l'EHESS, Paris.

Jullien, E. and Pauly, M. (eds) (2016). *Craftsmen and Guilds in the Medieval and Early Modern Periods*. Franz Steiner Verlag, Stuttgart.

Karpik, L. (1995). *Les avocats, entre l'Etat, le public et le marché, XIIIe-XXe siècle*. Gallimard, Paris.

Karpik, L. (2007). *L'économie des singularités*. Gallimard, Paris.

Larson, M.S. (1977). *The Rise of Professionalism. A Sociological Analysis*. University of California Press, Berkeley.

Le Bianic, T. and Vion, A. (eds) (2008). *Action publique et légitimités professionnelles*. LGDJ, Paris.

Malochet, G. (2004). À l'école de la détention : quelques aspects de la socialisation professionnelle des surveillants de prison. *Sociologie du travail*, 46(2), 168–186.

Maugeri, S. (2003). Une lecture pessimiste du processus de rationalisation. L'exemple des dispositifs de gestion logistique. In *Du politique dans les organisations*, Boussard, V. and Maugeri, S. (eds). L'Harmattan, Paris.

Menger, P.-M. (2002). *Portrait de l'artiste en travailleur. Métamorphoses du capitalisme*. Seuil, Paris.

Milburn, P., Kostulski, K., Salas, D. (2010). *Les procureurs : entre vocation judiciaire et fonctions politiques*. Presses Universitaires de France, Paris.

Monjardet, D. (1996). *Ce que fait la police. Sociologie de la force publique*. La Découverte, Paris.

Offerlé, M. (1998). *Sociologie des groupes d'intérêt*. Montchrestien, Paris.

Paicheler, G. (2010). Vivre ou évaluer le risque sanitaire. Expertises, controverses et confrontations profanes et professionnelles. In *Risques et pratiques médicales*. EHESP, Rennes.

Paillet, A. (2007). *Sauver la vie, donner la mort. Une sociologie de l'éthique en réanimation néonatale*. La Dispute, Paris.

Paradeise, C. (1984). La marine marchande française : un marché du travail fermé ? *Revue française de sociologie*, 25(3), 352–375.

Pichonaz, D. (2017). *Devenirs policiers. Une socialisation professionnelle en contrastes*. Antipodes, Lausanne.

Power, M. (1999). *Audit Society: Rituals of Verification*. Oxford University Press, New York.

Pudal, R. (2016). *Retour de flammes. Pompiers, des héros fatigués ?* La Découverte, Paris.

Riva, A. and Lagneau-Ymonet, P. (2012). *Histoire de la Bourse*. La Découverte, Paris.

Rivière, F. (2008). Guildes, monopoles et oligopoles dans la Normandie de la fin du Moyen Âge : la réglementation des métiers est-elle hostile à la concurrence ? *Entreprises et histoire*, 52, 36–45.

Rivière, F. (2017). Travail et métiers en Normandie à la fin du Moyen Âge : institutions professionnelles et régulation économique. PhD Thesis, EHESS, France.

Sewell, W.H. (1983). *Gens de métier et révolution. Le langage du travail de l'Ancien Régime à 1848. French translation*. Aubier-Montaigne, Paris.

de Terssac, G. (1992). *Autonomie dans le travail*. PUF, Paris.

Vigour, C. (2008). Ethos et légitimité professionnels à l'épreuve d'une approche managériale : le cas de la justice belge. *Sociologie du travail*, 50(1), 71–90.

Weber, M. (1905). *L'éthique protestante et l'esprit du capitalisme*. Librairie Plon, Paris.

Zarca, B. (2009). L'ethos professionnel des mathématiciens. *Revue française de sociologie*, 50, 351–384.

Zolesio, E. (2012). La chirurgie et sa matrice de socialisation professionnelle. *Sociologie*, 3, 377–394.

5

Professionalization as an Object of Tensions between Institutional, Collective and Individual Logics

Mokhtar KADDOURI
CIREL Laboratory, University of Lille, Villeneuve d'Ascq, France

After its battle against the qualification model, the competence model was subject to significant criticism regarding its social uses (Dugue 1999) and limitations. This gave rise to three major shifts: from careers to trajectories (Monchatre 2007); from skills to employability (Loufrani-Fedida and Saint-Germes 2013); and from training to professionalization (Jobert 2002, pp. 247–260). Starting from this last shift, we will explore the institutional, collective and individual logics underlying professionalization practices, as well as the tensional relationships between the actors involved in them. Sharing the theoretical option supported by Wittorski et al. (2015) as to the place of identity, skills, knowledge and action in professionalization processes, particular emphasis will be placed on the identity dimension of the latter. To do this, we will consider professionalization schemes as one of the expressions of the identity project that institutional leaders have with the groups or individuals to be professionalized, and the commitment of the latter to these schemes as one of the indicators of their own identity projects (Kaddouri 2005, pp. 145–157). We will also discuss that the incompatibility between the two identity projects is a source of tensions, which can play a driving role in professionalization processes.

Encyclopedia of Professionalization,
coordinated by Didier DEMAZIÈRE and Richard WITTORSKI. © ISTE Ltd 2024.

The first part of this contribution will delve into the status of the term "professionalization" which, through its polysemy and ambiguity, fuels, while simultaneously masking, the tensions between the actors of professionalization. The second part will address the tensional configurations of actors. These will be illustrated by an example drawn from one of our research projects. We will schematize such relationships by means of a professionalization polygon. The third part will study the relationships between and within the categories of actors and the logics underlying their relationships with the professionalization scheme. We will see that the strength of the tensions and conflicts which underlie the relations between actors is measured by the intensity of the challenges that professionalization represents for each of them. Professionalization that does not pose challenges to the actors is doomed to failure, or to unfinished attempts at rebirth, to say the least. The fourth part will rediscuss the identity dimensions of the professionalization scheme. We will analyze the three social uses that are made of it by three of the audiences in the professionalization polygon. Finally, the conclusion will return to some salient points with the intention of prolonging the debate regarding the challenges and potentialities, as well as the limitations of professionalization.

5.1. The ambiguous status of the term professionalization: between camouflage and the revelation of tensions between and within the configurations of actors

The polysemy of the term professionalization affects its theoretical and practical status, thereby confirming that behind designations lie challenges, purposes and goals which fuel tensions and conflicts between the actors, all the while camouflaging them. It authorizes differentiated social uses, with the risk of its overvaluation by some and its rejection by others. We know that when a term – or the thing it designates – is the object of desire or, on the contrary, of excessive rejection, it is because it is overloaded with meaning and constitutes a social challenge at a given moment in the history of a society, an organization, a social group and even an individual. Indeed, "words that convey strong social challenges are subject to underground attempts at capturing and shifting different meanings" (Mathey-Pierre and Bourdoncle 1995, p. 147).

Deleuze (1988) emphasizes that a concept is "that which prevents thought from being a simple opinion, a piece of advice, a discussion, chit-chat"[1]. So what about the term professionalization? Is it a category of analysis, or a category of practice

1 Quote drawn from a text by G. Deleuze entitled "*Signes et événements*" (1988), published in Véronique Delbove's blog (10/28/2005): http://laphilosuitsoncours.over-blog.com/article-909579.html.

(Brubaker 2001)? We share the viewpoint of Demazière (2009, p. 85) when he affirms that one of the difficulties of professionalization stems from the fact that it "is not so much a category constituted for the purposes of analysis and understanding of the world of labor, but a category with a practical and singularly polysemic aim". There have been numerous attempts to elevate it to the rank of a concept and the combined efforts have succeeded in making it an "operational" concept, rather than a systemic one (Bourdieu 1968). But despite, or perhaps thanks to, its polysemy, its rhetoric – as it happens with the term "skill" – has invaded all of the fields of professional practices, where it occupies a central place in employment policies, work organization and the training function (Sorel 2015, pp. 189–217). In these fields, the term designates different facets of the same phenomenon and, where definition conflicts could contribute to clarify the plurality of its meanings, the term is invoked as an instrumental strategy to justify the aims and goals of an actor (individual or collective). And it is this possibility of the term to designate the different dimensions of a phenomenon (or the same phenomenon considered in different ways) that enables it to serve several strategies and goals at the same time, by endowing it not only with virtues, but also with the dangers of "functional ambiguity", as in the famous expression coined by Durkheim (1912, p. 367). Speaking of the pure and the impure, the author (ibid., p. 389) signals the presence of two kinds of sacred:

> one prosperous, the other harmful, and not only is there no solution of continuity between the two opposing forms, but the same object can pass from one to the other without changing its nature. With the pure, we make the impure, and the other way around. It is in the possibility of these transmutations that the ambiguity of the sacred is found.

Paraphrasing him, it is in the possibility of transmutations that the polysemy of the term professionalization allows that lies its ambiguity. This endows it with a remarkable plasticity, which allows it to serve different functions to the satisfaction of several actors who, despite their divergent interests, succeed in establishing circumstantial alliances, which do not prevent them from pursuing their own strategies separately, and sometimes, from hiding their real intentions. For example, this is the case when we are in the presence of actors who seek to "artificially silence the differences between training schemes that are nonetheless differentiable [...]" (Broise 2012, p. 135) by making them pass for something they are not, to the point that we can join Wittorski (2007 p. 11) in his question: "nowadays, what training scheme does not claim to be aimed at professionalization?".

We have already encountered this dissimulation in the context of adult training, cluttered with practices having nothing to do with training, and yet designated as such and capitalized in a fallacious manner, by certain companies as part of their training actions. This pushed the legislative authority to establish definitive rules and indicators, thus preventing any form of amalgamation. Professionalization runs the same risk if (in the name of skills development or any other similar pretext) any action taken is characterized in this way. Wishing at all costs to attribute this qualifier to an action which it is not amounts to seeking to make it fulfill functions, which elude its real aims, attempting to justify it as legitimate. We can clearly see it: trapped in its "functional ambiguity", the notion of professionalization enables each actor to pursue their own goals and their own purposes. Referring to the theory of translation, Poumay et al. (2017, p. 7) recall "that a concept is all the more interesting when the actors can both translate it and betray it in problematizations and singular actions". It is these possibilities of its differentiated uses that makes the notion of professionalization applicable to multiple objects, endowing it with the gift of ubiquity, which creates "the temptation to see the world only through its lens" (La Soudière 2000, p. 5). But this functional ambiguity, the gift of ubiquity and the betrayals it enables do not stand up to the test of reality, especially when the consequences of the professionalization process disrupt the temporary and provisional balance between its actors, insofar as ambiguity is tested by the tensions that agitate them. After these clarifications, let us now return to the tensional relationships between the different actors of professionalization. These will be analyzed in the light of an example drawn from our previous research.

5.2. Professionalization as a configuration of actors engaged in interdependent and tensional dynamics: illustration of configurations of actors in tension around a professionalization scheme

As any social practice, professionalization is affected by numerous challenges that may be economic (employment, productivity, protection against competition, etc.), socioprofessional (staff flexibility, adaptation to work organization, adherence to new managerial methods and technical-industrial projects, etc.) or related to socialization and identity positioning (recognition, attribution or claims to having a certain social status; legitimization of an identity; etc.). These different challenges have no ontological existence, but are located within the frame of social and power relations which underlie the positions of the actors in relation to them. This is not presented "aboveground", but rooted in an institutional and organizational framework which structures its goals, purposes, allocated resources, the public benefiting from it and the roles system of the actors it engages. To a large extent, these are the same challenges encountered in the context of the illustrative example below.

After having retraced the context of the industrial project, we will return to the different actors and identify their tensional relationships, by schematizing them in a polygon we have come to call the "professionalization polygon".

5.2.1. *The context of the professionalization project*

The examples discussed herein are drawn from research in an establishment, which is part of a French group in the agri-food sector, specializing in the slaughter and processing of pork and beef. To face competition and national and European competitiveness, the management decided to introduce profound changes. The old buildings were destroyed and new, sleeker and more modern buildings were constructed. The entire industrial apparatus used until then was abandoned, shifting toward a new technology for the generalization of computerized management and production. Formerly, the company had a dual meat processing business. The new company shifted to a single-product economic activity: the slaughter and processing of pork, which, consequently, led to the abandonment of beef activities. The factory's organization was designed in the form of "mini-companies", each having a certain management autonomy. From that moment, the production circuit was controlled downstream (the market), instead of upstream (by breeders and livestock suppliers). Due to political and institutional reasons, the management decided to keep all staff (400 employees) by setting up a professionalization scheme to prepare them for the sociotechnical and organizational changes envisaged.

It is in this context that our research took place (Kaddouri 1992), as part of a larger research project conducted by a multidisciplinary team which will be discussed further. To collect our data, we cross-referenced several tools: documentary analysis (social reports, works council reports, training sheets, etc.) and semi-structured interviews with 30 employees organized around five themes (biographical data, representations of the envisaged change; representations of the professional and personal future in the upcoming company, projects and strategies for its achievement, attitudes toward training). Interviews with company managers (the company director, the human resources manager, the sales director, the two managers of the cattle and pork sectors) essentially focused on the change components, the expected (or feared) effects, the purposes and goals of the professionalization scheme. Furthermore, we observed some training actions preparing the changes.

The content analysis of the information reveals the existence of a dual intention on the part of the company's management: to prepare the agents to acquire the technical skills deemed necessary for the success of the transformations envisaged and to produce, individually and collectively, a new professional identity structured around the processing of pork. These two intentions structured the content of the professionalization scheme. The latter engaged seven stakeholders: the company management, the engineering firm, the public authorities, the company staff, the staff representatives, the village residents and the research team.

5.2.2. *Actors in tension around the professionalization scheme*

We rely on the idea of "linked ecologies" theorized by Abbott (2003, pp. 29–50) for the presentation of the seven audiences above. The idea of "linked ecologies" encompasses the relationships between the actors in both their intra- and inter-dynamics, taking into account, on the one hand, the "possibilities, even probabilities, for certain alliances to be established even beyond the boundaries of these ecologies, [...]", and on the other hand, the unfolding of "interactions between ecologies", [...] "the conflicts, the bargaining and the negotiations which lead to the alliances" , as well as the "hinges", and "articulations between the operations of the two ecologies" (Abbott 2003, p. 16, paragraph 67). Depending on the interests that motivate them, each of the seven audiences favored one or the other dimension of the professionalization scheme, bringing it closer to certain actors while opposing it to others.

The main challenges for the company's management – related to the success of the industrial project – were the use of professionalization scheme to strengthen the spirit of entrepreneurship, the acceptance of technical and organizational changes by all staff and the consolidation of an organizational culture to the detriment of a strong métier culture, especially in the cattle sector. For the success of its industrial project, the management needed not only subsidies from the public authorities (and partly from the European Commission), but also their support and project legitimization. The on-site presence of an engineering firm, the representatives from the French Ministry of Agriculture and a multidisciplinary team of researchers could, in their eyes, only reinforce such legitimacy. Researchers were expected to provide a better understanding of staff concerns, of their representations and attitudes toward changes, with a view to readjusting the professionalization scheme and obtaining their commitment to the goals, culture and challenges of future business.

The engineering firm, responsible for the implementation of the industrial project, played a dual role: as a contracting authority of the professionalization project and as a training operator. Its challenges concerned the strengthening of its legitimacy within the company and the acceptance of the professionalization scheme by the staff and their union representatives. Another challenge involved gaining recognition in other fields of activity by being accepted as competent to lead and to support the implementation of projects of this nature and of such magnitude. In fact, it had strong bonds with the management and aimed to strengthen them with the researchers with a dual purpose: to gain feedback on the progress of the professionalization project as expressed by the employees during interviews, and to benefit from any piece of advice that would help them to adjust elements throughout the professionalization process.

The public authorities were involved in the project through the commitment of the French Ministry of Labor (professional training delegation and ANACT (*Agence nationale pour l'amélioration des conditions de travail*)) and the French Ministry of Agriculture. It was also through the General Directorate of Education and Research (*Direction générale de l'enseignement et de la recherche*) of this ministry that the researchers were put in contact with the company. One of the major concerns for the public authorities, in connection with the professionalization scheme, was to keep employment within the company and failing that, to prepare professional mobility within the frame of the industrial group. This was one of the conditions for granting subsidies to the company. One of the challenges for representatives of the French Ministry of Agriculture was the transfer of experience of change and professionalization to the agricultural sector. This explains the interest they placed on the research results.

Special emphasis was given to the company's staff in the cattle and pork sectors. The first, taken here as "ordinary employees outside any regulated group" (Dubar et al. 2011, p. 306), were in favor of the professionalization scheme. For them, getting involved meant being part of the new company, a more than worrying prospect. Staff in the cattle sector were fiercely hostile to the technical-organizational change project, deemed threatening to their survival as a professional group in the sense of Demazière and Gadéa[2] (2009, p. 20) and Demailly[3] (2008, p. 79). Under these conditions, engaging in the professionalization scheme would amount to supporting the reconversion of their identity, conveying a degradation of their professional status within the company. This would also amount to swapping a rewarding métier (beef is a noble material) for a degrading one, since pork is considered to be a "dirty material", requiring no particular skills. On the contrary, cutting beef requires a special qualification (CAP or certification of professional aptitude) in butchery. For

2 Demazière and Gadéa (2009, p. 20): "groups of workers carrying out an activity with the same name, endowed with social visibility, benefiting from a social identification of work, and characterized by symbolic legitimacy".

3 Demailly (2008, p. 79) proposes three criteria for identifying them: "a similar and relatively stable place in the division of labor, extra-organizational and extra-territorial [...]; a certain specific and collective subjective existence; a certain specific and collective capacity for intervention in the social field".

all of these reasons, staff in the cattle sector were fiercely opposed to the company's management and those who supported its industrial project, notably the engineering firm. Some saw the research team as a possible spokesperson for their demands and fears. The staff delegates and some of the village residents were valuable allies.

Staff representatives: in the past, there was a CGT (*Confédération générale du travail*) union section which was created in 1968, following a movement which enabled it to obtain the transition between two forms of minimum wages: from the SMAG (*Salaire minimum agricole garanti*) to the SMIG (*Salaire minimum interprofessionnel garanti*). But gradually the section disappeared due to several factors: failure of a second strike carried out in 1975, pressure from the management, the retirement of a pillar activist of the section, etc. Concomitantly with this disappearance, nonunion elected officials became heads of the staff committee, whose activity was reduced to the organization of cultural and sports. The members of the CHSCT (*Comité d'hygiène, de sécurité et des conditions de travail* (Committee on Health, Safety and Working Conditions)) were consulted on minor problems regarding future operation. The two CGT union delegates who managed to maintain their position did not benefit from sufficient balance of power to influence the process of change, nor the professionalization scheme which prepared for it. The support they provided to staff in the cattle sector was more symbolic than effective in reality.

The village inhabitants, some of whom joined together in breeder cooperatives, were not only affected by the disappearance of the cattle sector, but also indirectly by the professionalization scheme. Beef breeders lost a local outlet for their livestock and families where one member worked in the cattle sector feared unemployment, or the obligation for one of their members to move geographically. The disappearance of beef was as much a socioeconomic blow as symbolic one to the village which dedicated an annual festival to it, "the beef festival", a source of pride for the villagers.

The research team: beyond the concerns specific to each discipline represented (sociology, ergonomics and education and training sciences), the shared preoccupation regarding the professionalization scheme revolved around the analysis of professional identities, the restructuring of professional groups (namely, the cattle and pork teams), and the relationship to work, change and the professionalization scheme proposed (or rather imposed) on all company staff. PIRTTEM[4] CNRS was added to the research team itself. Its representative

---

4 Interdisciplinary Research Program on Technologies, Work, Employment and Lifestyles (*Programme Interdisciplinaire de Recherche sur les technologies, le Travail, l'Emploi et les Modes de vie*).

accompanied the research team throughout the period of its presence within the company. The researchers' challenges were of an epistemic nature, on the one hand, pertaining to the production of knowledge regarding the objects of research and the orientation of their discipline. On the other hand, they were of an epistemological order, particularly related to the conditions of knowledge production in a context where the risk of exploitation by the actors of the project weighed on them (both those who led it and those who suffered from it). The challenge at this level was not to be perceived as a stakeholder in the process while conducting the professionalization project and/or to be taken as a spokesperson for the opponents of industrial and/or professionalization projects.

After this illustrative presentation, we have now schematized a hexagon that locates the seven audiences and the challenges, having mobilized[5] them around the professionalization scheme.

First of all, remember that a hexagon is a polygon[6] whose angles and sides (as well as the lines which connect them) may vary in length and amplitude depending on the configurations, contexts and situations in which the actors of professionalization work. The idea of the angles makes it possible to grasp the places from which these actors express their viewpoints, justify their policies and practices in this area, whereas the sides refer to the different dimensions of professionalization and indicate those requiring the attention of a specific actor in the polygon, be them of a political, organizational, activist, managerial or educational nature. The lines that connect the actors allow us to map their interrelationships (or lack of them). So, for each actor, we have an angle, a side and lines that connect it to the other actors in the polygon. Due to their double or triple positioning, some actors can express themselves from two or three angles and sides. This poses problems in terms of role conflicts when an actor intervenes as contracting authority in a professionalization scheme, while being one of the operators at the same time. This is what happened with the engineering firm.

5 The interested reader can use the table below to construct a professionalization polygon:

Actors of professionalization	Goals	Challenges	Alliances and oppositions

6 A polygon is a geometric shape bearing different names, depending on the number of its sides: triangle (3 sides), quadrilateral (4), pentagon (5), hexagon (6), heptagon (7), octagon (8), etc.

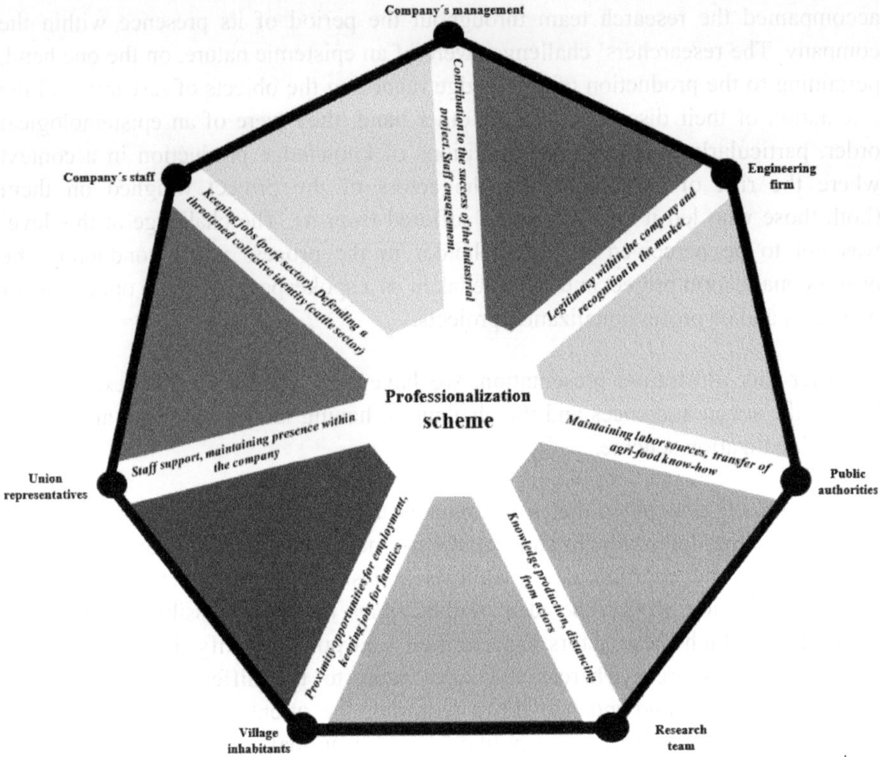

Figure 5.1. *Professionalization polygon. For a color version of this figure, see www.iste.co.uk/demaziere/encyclopedia.zip*

5.3. Conflicting relationships and reconfigurations of variable actors around the professionalization scheme

As we have just seen, far from being mechanical, the relationships between the different actors in the hexagon are dynamic, because they are crossed by social relationships; they are interdependent and interactive, mobilized by challenges around the same object (here, a professionalization scheme preparatory to a radical technical-organizational change). These relationships are not contemporary with the decision for the scheduled change, but part of sociohistorical and normative frameworks that structure such relationships. While the staff majority come from the local families, management comes from the ranks. The company, created from scratch in 1961 by local actors, experienced events that marked its current social relationships (strikes, family ties in the village extending within the company, annual beef festival, etc.). The elements below will serve as a reading grid for the relationships between the categories of actors and the logics that underlie their positioning in the professionalization hexagon.

5.3.1. *Some characteristics of the relationships between and within the categories of actors*

Asymmetrical and unequal relationships within the frame of contingent and evolving configurations. First of all, it should be noted that depending on their position within the company and being aware of the balance of power which unites them, the challenges fueling them and the goals they pursue, these actors maintain unequal and asymmetrical relationships toward professionalization. These are unequal relationships in that they do not wield the same resources and are not endowed with intellectual, material and symbolic capital which are likely to influence the sense and direction to be given to professionalization. These relationships are also asymmetrical insofar as the social and professional positions that the actors hold within the company do not enable them to experience the challenges and effects of the envisaged change and the professionalization scheme preparing it in a balanced manner. In fact, the positions within the hexagon have a variable-geometry and the intensity of the actors' investment and engagement vary depending on the direction, which structures their logic of action and the importance they give to the professionalization process. Thus, the cattle sector staff, the union delegates and the village inhabitants, on the one hand, and the company's management, the engineering firm, on the other hand, did not share the same positioning in relation to the professionalization scheme. The former were undergoing a change they did not want; quite the opposite, their attachment to what existed was stronger than a temptation to embrace the change envisaged by the management. The latter, supported by the engineering firm and conditionally legitimized in its project by the public authorities, intensified its bet on the professionalization scheme. While the different audiences undoubtedly intensified their commitment to the professionalization scheme (either by supporting or rejecting it), it was the asymmetry and inequality that characterized their relationships to it.

Tensions within the categories of actors: even if the actors mentioned have been introduced in the form of homogeneous categories, this should not hide their own internal contradictions and tensions. This is the case of the state, as Abbott (2003, pp. 29–50) observes, featuring it as one of the important actors of professionalization. According to this author, the state "is not a unified audience, but a scheme of small groups waging their own little wars for their own little reasons" (p. 4, paragraph 15). The same goes for each of the seven audiences which, taken in isolation, can be shaken by tensions, even internal conflicts which are sometimes latent and at other times declared, thereby calling into question the internal arrangements pertaining to the process conducting and guiding the direction of professionalization. Let us discuss two examples. The first concerns the tensions within the personal audience of the cattle sector, which suggested strong internal cohesion. However, there was an opposition between those who refused the project in its entirety and did not want to integrate the new company at any price, and those who, due to a lack of resources and a clearly unfavorable balance of power, reluctantly agreed to join the pork sector and thereby, the professionalization scheme. Several scenarios were observed within this audience: those who set up on their own as a butcher-artisan, external or internal professional mobility within the group's site having a cattle sector or retirement. On the other hand, those who lacked the resources to go elsewhere agreed to mourn their bovine identity and integrate into the pork sector. They agreed to join the professionalization scheme to "secure a place" and "save their skin" in the new company. Here, we have another type of use of a professionalization scheme: its use to protect themselves from the consequences of a change one fiercely refuses.

The second example of tensions was to be found at the heart of the research team audience, torn between divergences as to the status of research and the relationship with institutional control, the company's management and the engineering firm. In concrete terms, divergence arose as to the role of researchers within the company. For the ANACT, it was a question of holding a position as intervening researcher and of being able to give "advice" to the management and the engineering firm, especially regarding the results of the analysis of working stations. The rest of the researchers advocated a research posture limited to the analysis and interpretation of the data collected, without providing any kind of "advice" to the management or the engineering firm. On the other hand, intermediate feedback sessions were planned to present the results of the research. The company's management and the engineering firm were free to appropriate them and translate them as part of their change management approach and in the implementation of the professionalization scheme.

Inter-category alliances beyond the tensions shaking them. The example also shows the existence of tensions between categories. This is what we observed, in the case of the interrelations between the company's management and the engineering, firm on the one hand, and the cattle sector staff, on the other hand. Frankly opposed to the change project and the professionalization scheme, the cattle sector staff knew that the change would cause their work tools to disappear and undermine their collective professional identity. It also suspected that the professionalization scheme constituted a "manipulative strategy" on the part of the management and the engineering firm to make employees accept the disappearance of their identity. As a result of this collective diagnosis, it organized passive resistance (no collective action at the time of the field survey). Due to an unfavorable balance of power, the cattle sector staff chose to establish alliances, even if contingent and provisional, so as to consolidate their positioning, and strengthen their balance of power and leeway to guide professionalization, in the direction of their own interests (in this case, trying to prevent its implementation). In its conflict with the company's management, the cattle sector staff sought to compensate for the unequal and asymmetrical relationships in relation to professionalization processes through alliances likely to strengthen their positioning and make the sector's resistance and demands audible.

5.3.2. *Multiple logics in tension*

The example also allows us to become aware of the existence of divergent institutional, collective and/or individual logics. By *institutional logics*, we mean "the systems of rules, tools and values associated with institutional actors, which provide lines of action and the understanding of situations" (Charue-Duboc and Raulet-Croset 2014, p. 30). Paraphrasing Amblard et al. (1996, p.199), the *actor logics* are defined here as "choices made by the individual or collective actor" whose analysis makes it possible to "account for what underlies them". The term *tension* refers to "the underlying idea of opposing forces, of actors (or groups of actors) who collide or even fight for resources, values or goals" (Guedri et al. 2014, p.14). The crossover between *tensions* and *actor logics* helps us understand that, far from being bad faith, the tensions between actors are driven by different logics of action at work in their relationships with regard to the professionalization scheme. We have identified three types as follows.

First of all, tensions between economic logic and social logic. The first is driven by company managers who are concerned about their economic positioning on competing national and European markets. The social logic is driven by the public authorities and their representatives, as part of the change project. Let us recall that one of the conditions for the subsidies to be granted by the state to the company was the safeguarding of employment, that is to say, there was a ban on layoffs. This logic was supported by the union delegates, by the staff and the village inhabitants. But this tension between the economic logic and the social logic gradually softened, insofar as the various actors involved in driving technological change became aware of the importance of the social component for attaining productivity goals and economic gain.

Tensions between the organizational logic and métier's logic. The organizational logic is observable through the new directions taken by the company's management, namely the transformation of production processes by focusing on a single activity (only processing pork meat). This transformation was accompanied by a technical and social organization of work into "mini-companies" and the strengthening of procedures downstream to the detriment of upstream tasks. This new organizational logic signified a heavy blow to the métier's logic, which had previously prevailed within the company. It was particularly fatal for the cattle sector staff. This called into question its qualifications, its acquired skills and its autonomy of operation. But this tension did not lead to collective action: to "activities through which a group seeks to act on the arrangement of the social order and to promote its claims" (Bloess et al. 1997, p. 12). Lacking in balance of power, and in the words of Dubar (1991), we come upon a professional group which did not succeed its "relational transaction" and found itself in a "blocked identity".

Tensions between the collective logic and individual logic. This tension was manifested on two levels. First of all, between the cattle sector staff acting collectively to defend their métier's identity, and the pork sector staff who seemed to have approached change in a rather individualistic manner. While the former refused change by joining collective resistance to defend their space and working tools, the latter did not show any collective resistance. Then, the logics tension exploded to the interior the cattle sector staff, torn between those who refused any transfer to the pork sector and those who, at a given moment, began to subscribe to individual logics (leaving, setting up on their own, joining the pork sector, etc.). Confronted with the impossibility of imposing defense and therefore the survival and preservation of their collective identity, the cattle sector staff made individual choices, giving rise to possibilities for differentiated positioning: leaving, setting up their own businesses, sticking to the beef sector, converting to the pork sector, etc.

So far, we have focused on our illustrative example to identify the logic of the different audiences and their tensional relationships within the frame of a localized professionalization scheme. Let us now return, in the last part of this contribution, to the identity dimensions of the schemes in question.

5.4. Professionalization as the production of skills and of professional identities

Considered as one of the major tools for professional socialization and the construction of identities at work, professionalization is characterized by a triple use: institutional, collective and individual. As we will see, in these three uses, professionalization is characterized by the existence of two levels of intentions:

skills acquisition and the construction of a professional and social identity. These are the two levels that we encountered among the three actors of the professionalization polygon: the company's management, the professional group making up the cattle sector staff and the employees (those from the pork sector taken individually).

5.4.1. *Professionalization as an institutional offer of skills and identities*

Seeking to professionalize their employees, employers implement professionalization schemes with a dual intention, and therefore a dual use. The first one involves skills acquisition as one of the dimensions of professionalization, with the aim of optimizing the productive performance. The second use concerns its identity dimension, that is to say professional socialization and the production of a professional identity which conveys a culture, a posture, a positioning consistent with the goals and purposes pursued by the managers of the organizations concerned. The challenge here is to get employees to agree to engage not only their skills but also (and above all) their subjectivity at work. These two professionalization levels are closely intertwined and reinforce one another.

5.4.2. *Professionalization as a claim to skills and collective professional identity*

In the second use of professionalization, we find the presence of an officially organized professional group (or considered as such), which also has a double intention: to professionalize itself as a form of protection from the threatening competition of other professional groups, and to position itself on its (internal or external) professional market and defend its autonomy at work and within its own territory. The goal is also to forge a recognized and legitimized collective identity, the recognition of a professional and social position and status, prestige and professional and social usefulness.

5.4.3. *Professionalization as a factory of skills and an individual professional identity*

In the third use, we find ourselves, on the contrary, in the presence of an individual actor who decides to professionalize itself by mobilizing a set of specific approaches, or by engaging in the professionalization measures proposed. Here too, we can consider engagement in a professionalization scheme as driven by a double intention: the construction of professionality, that is to say, the "knowledge and abilities used in professional practice" (Bourdoncle 1991, p. 76) and the construction of a professional identity, a posture, a positioning, a culture consistent with the professional values and ethics which underlie a specific vision of the métier practiced or envisaged. This consistency constitutes a major motivation, encouraging employees to engage their skills at the service of productive performance.

Now the question which arises is: what relationship can be established between the supply and demand for professionalization, that is to say, between the professional and identity project for the others (institutional use) and the professional and identity project for ourselves (second and third use)?

Relationship between the project for ourselves and self-project for the others

Empirical observation shows that these identity relationships are often in tension and crystallize around the role of professionality in the transactions between the employers (who seek a finalized mobilization of skills and salary identities with a view to optimizing job gains) and the employees (who seek to have their skills and identities recognized, both individually and as a professional group). Table 5.1 will allow us to summarize the two professionalization levels as well as the triple use made of it.

Professionalization levels → Audiences ↓	First level: skills	Second level: identity postures
Employers	Acquisition of skills required for productive performance.	The inculcation of a positioning and posture enabling the mobilization of salary subjectivities at work.
Professional group	Skills acquisition which protect against competition and help position ourselves in the market by defending our autonomy at work and our own territory where to exert it.	Recognition of a collective professional identity, a professional and social position and status, prestige and professional and social utility.
Employees	Skills acquisition and construction of our own professionalism.	The construction and/or consolidation of a professional identity aimed at the recognition of a desired professional status.

Table 5.1. *Professionalization levels and their uses*

Reading the table vertically (by column) allows us to visualize the purpose of the first level of professionalization from the side of the employer, as well as that of the professional group and the employee. This concerns skills acquisition, but for three different goals. For the employer, skills acquisition aims at productive performance, for the professional group, the defense of a market and a professional territory, whereas for the employee, it aims at building their own professionalism. This vertical reading also makes it possible to visualize the object of the second level of professionalization: the acquisition of a posture and an identity positioning. Here too, acquisition is used for achieving three different goals. In one case (the employer), this involves subjective mobilization at work with a view to increasing productive performance. For the professional group, it strengthens the legitimacy of a status and a collective identity. For the employee, it is a claim for obtaining, from the employer, recognition and professional status.

The horizontal reading of the table (by rows) allows us to visualize the two professionalization levels, namely the skills and the identity posture, as much for the employer (first row) as for the professional group (second row) and the employee (third row). It allows us to see that the employer, through the link established between the mobilization of skills and salary subjectivity at work, is indeed aiming for productive performance. For the professional group, it is a question of situating itself in its market or professional territory where the métier is practiced and having the collective identity and socioprofessional status resulting from it. For the employee, the link established between skills and the mobilization of their subjectivity at work aims at the recognition of their individual identity and the professional status associated with it.

5.5. Conclusion

Professionalization can be understood as an element within a "system of interactions" subject to power relations. We have seen it throughout this contribution, professionalization is not only a question of engaging skills and resources to optimize productive performance and build an identity positioning. It is more than that, since, as Bronckart (2009) suggests, it involves an engagement which is "only a secondary aspect of a more global mechanism permanently reorganizing three types of relationships" (the relationship to the situation, to others and to ourselves). These three types of relationships are underpinned by social challenges and power relations in which persons seeking professionalization and the institutions offering it are not isolated in "an ivory tower", far from the social challenges and power relations which connect them to other "interaction schemes", in the sense of Hugues (1996). Paraphrasing him, we could say that professionalization, as a social practice, can only be made intelligible within the frame of a "social matrix" and the "social scheme" it is part of (ibid., 1996, p. 66). This is what we observed in our example. The sense that the professionalization scheme took on for the various actors present could only be understood within the frame of the technical-organizational change project envisaged by the management. This was at the origin of the disruption of the interaction systems and the "social matrix", which structured the interrelational balance between the different actors of the former company. It was an opportunity for success for some, and a factor of risk and threat for others. Thus, the sales department staff saw their positioning strengthened to the detriment of the production department staff. This resulted from the transition from production dominated upstream (breeders and suppliers of beef and pork) to production dominated downstream (the company only producing what is already sold on the market by the sales department). The cattle sector staff proved to be the big losers from such change. It disintegrated as a professional group and its collective identity vanished, giving rise to individual survival strategies, since it was unable to build a balance of power to impose its collective future in the upcoming company.

5.5.1. *Professionalization as a tensional combination of interdependent actor logics*

In the introduction, we hypothesized a structuring role of tensions in professionalization processes. The diversity of strategies and the goals they serve, the inequality and asymmetry of relationships with regard to the balance of power and positioning within the company, led each actor in the polygon to attempt to reconfigure the professionalization scheme according to its own logic. As we will recall, apart from a few strategic positions requiring technical expertise and advanced skills, this constituted a sort of "cultural revolution", as it replaced a new technical and social organization of work (mini-companies, etc.) while introducing unprecedented relationships with the environment (management of production downstream to the detriment of upstream). It was crossed by a diversity of logics of actors: economic logics and social logics; organizational logics and the métier's logics; and collective logics and individual logics. Without readdressing each of these logics whose content has been previously explained, we recall here, by way of illustration, the tensional relationship between collective logics and individual logics having given rise to differentiated relationships with regard to the professionalization scheme. Following an individualistic logic, by committing massively to the professionalization scheme, the pork sector staff had the goal of "booking" their place in the new company. By its encouragement and valorization of this commitment, the company's management gave the confirmation of being taken and renewed in the future company. The cattle sector staff expressed a position of radical hostility, judging that it marked their collective disappearance, as it prepared them for a biographical and identity transition in which they found themselves powerless to succeed (Deltand and Perez-Roux 2017). Part of a collective logic and refusing to engage in the professionalization scheme, the cattle sector staff demonstrated their resistance and their refusal to see their collective professional identity disappear.

5.5.2. *Resistance to professionalization as a factor asserting our professionalism and claiming identity*

We learn from our example that becoming professional involves not only confirming or seeking to acquire new skills, but also of resisting the loss of existing professionalism. Thus, by offering the cattle sector staff to engage, in the same way as other staff members, in its professionalization scheme, the company's management thought it was doing the right thing by seeking to prepare them for the envisaged change. But it was different for the staff in question, for whom the professionalization scheme was not only a sign of their degradation and their predicted death but also the prelude to their deprofessionalization process. Here, as Monceau (2012, p. 115) points out "the professionalization of métiers is worked through the negative pathway. It is also a form of deprofessionalization in the sense that it generates the withering away of the métier(s) it comes from". Among the cattle sector staff, the feeling of deprofessionalization was expressed through the attack to their professional status, their place and their praised position within the former company. In this context, the sector's resistance to the feared deprofessionalization illustrates, one of the forms of resistance to the disappearance of a collective professional identity, and the rejection of an identity which was sent back to them as in a distorting mirror, by the established analogy between the cattle sector and the pork sector: nobility of the métier, on the one hand, and dirty and degrading job, on the other hand. If we adopt the viewpoint of the cattle sector staff, we can only share the following question by Edgar Morin (1979): "do not professionalization processes which appear to us as processes of liberation, protection and salvation include a dark element of enslavement and perdition?". This is the prevailing feeling that the professionalization scheme generated among the cattle sector staff.

5.6. References

Abbott, A. (2003). Écologies liées. In *Les professions et leurs sociologies*, Menger, P.-M. (ed.). Éditions de la MSH, Paris.

Amblard, H., Bernoux, P., Herreros, G., Livian, Y-F. (1996). *Les nouvelles approches sociologiques des organisations*. Éditions du Seuil, Paris.

Bourdieu, P. (1968). *Le métier de sociologue*. Mouton/Bordas, Paris.

Bourdoncle, R. (1991). La professionnalisation des enseignants : analyse sociologiques anglaises et américaines. *Revue française de pédagogie*, 94, 73–92.

Boutang, P.-A., Deleuze, G., Parnet, C. (1996). *L'abécédaire de Gilles Deleuze*. Éditions Montparnasse, Paris.

Broise, P. (2012). La professionnalisation au risque de la norme ? In *La professionnalisation mise en objet*, Demazière, D., Roquet, P., Wittorski, R. (eds). L'Harmattan, Paris.

Bronckart, J.-P. (2009). *La notion de compétences est-elle pertinente en éducation ?* L'éducation démocratique, Paris.

Brubaker, R. (2001). Au-delà de l'identité. *Actes de la recherche en sciences sociales*, 4(139), 66–85.

Charue-Duboc, F. and Raulet-Croset, N. (2014). Confrontation de logiques institutionnelles et dynamique des routines organisationnelles. *Revue française de gestion*, 240, 29–44.

Deltand, M. and Perez-Roux, T. (2017). Les dynamiques identitaires à l'épreuve des transitions. *Education permanente*, 212, 5–14.

Demailly, L. (2008). *Politiques de la relation. Approche sociologique des métiers et activités professionnelles relationnelles*. Éditions Septentrion, Lille.

Demazière, D. (2009). Professionnalisations problématiques et problématiques de la professionnalisation. *Formation Emploi*, 108, 83–90.

Demazière, D. and Gadéa, C. (2009). *Sociologie des groupes professionnels. Acquis récents et nouveaux défis*. La Découverte, Paris.

Dubar, C. (1991). *La socialisation. Construction des identités sociales et professionnelles*. Armand Colin, Paris.

Dubar, C., Tripier, P., Boussard, V. (2011). *Sociologie des professions*. Armand Colin, Paris.

Dugue, E. (1999). La logique de la compétence : le retour du passé. *Education permanente*, 140, 7–18.

Durkheim, E. (1912 [1968]). *Les formes élémentaires de la vie religieuse. Le système totémique en Australie*. Presses Universitaires de France, Paris.

Étienne, J., Bloess, F., Noreck, J.-P., Roux, R. (1997). *Dictionnaire de sociologie, les notions, les mécanismes et les auteurs*. Éditions Hatier, Paris.

Guedri, Z., Hussler, C., Loubaresse, E. (2014). Contradictions, paradoxes et tensions en contexte(s). *Revue française de gestion*, 3/240, 13–28.

Hugues, E. (1996). *Le regard sociologique. Essais choisis. Textes rassemblés et présentés par J-M. Chapoulie*. Éditions de l'École des Hautes Études en Sciences Sociales, Paris.

Jobert, G. (2002). La professionnalisation, entre compétence et reconnaissance sociale. In *Formateurs d'enseignants. Quelle professionnalisation ?* Altet, M. (ed.). De Boeck Supérieur, Louvain.

Kaddouri, M. (1992). La gestion du changement : actions et processus impulsés par l'entreprise. *Études et expérimentation en formation continue*, 14, 19–22.

Kaddouri, M. (2005). Professionnalisation et dynamiques identitaires. In *La professionnalisation en actes et en questions*, Sorel, M. and Wittorski, R. (eds). L'Harmattan, Paris.

La Soudière, M. (2000). Le paradigme du passage. *Communications*, 70, 5–31.

Loufrani-Fedida, S. and Saint-Germes, E. (2013). Compétences individuelles et employabilité : essai de clarification de leur articultation. *Association de gestion des ressources humaines*, 7, 13–40.

Mathey-Pierre, C. and Bourdoncle, R. (1995). Autour du mot "professionnalité". *Recherche & Formation*, 19, 137–148.

Monceau, G. (2012). Le dilemme professionnel : mise en tension de la pratique professionnelle par la professionnalisation. In *La professionnalisation mise en objet*, Demazière, D., Roquet, P., Wittorski, R. (eds). L'Harmattan, Paris.

Monchatre, S. (2007). Des carrières aux parcours... en passant par la compétence. *Sociologie du travail*, 49, 514–530.

Morin, E. (1979). A propos de la déprofessionnalisation. *Critère* [Online]. Available at: http://agora.qc.ca/documents/Profession–A_propos_de_la_deprofessionnalisation_par_Edgar_Morin.

Poumay, M., Tardif, J., Georges, F. (2017). Introduction. In *Organiser la formation à partir des compétences : un pari gagnant pour l'apprentissage dans le supérieur*, Poumay, M. (ed.). De Boeck Supérieur, Louvain.

Sorel, M. (2015). La professionnalisation, un processus en tension. In *Les professionnels et leurs formations*, Wittorski, R., Maulini, O., Sorel, M. (eds). Peter Lang, Berne.

Wittorski, R. (2007). *Professionnalisation et développement professionnel*. L'Harmattan, Paris.

Wittorski, R., Maulini, O., Sorel, M. (2015). *Les professionnels et leurs formations. Entre développement des sujets et projets des institutions*. Peter Lang, Bern.

PART 3

Professionalization, the Construction of Professionalities

6

Apprenticeship in Education and in Training: Foundation for Adult Training, Professionalization Analyzer of Educational Paths and Pedagogic Figure

Philippe MAUBANT
CRIFPE, University of Sherbrooke, Canada

The text will begin by a reflection on the conditions for the emergence of apprenticeship in the dual field: school education and adult training. Starting with an analysis of the foundations of the discourse of institutions and organizations, these first pages help identify the ideologies and utopias at the origin of the invocation of the concept of apprenticeship. To begin this first section, the concept of apprenticeship is put into dialogue with the question of temporalities and spaces. Studying apprenticeship through the question of temporalities will also lead us to raise the question of spaces; in other words, the environments where apprenticeship takes place.

The second section of the chapter proposes putting the use of the concept of apprenticeship into perspective, in the light of strategic developments having an impact on work and training organizations. Nowadays, the injunction to professionalize métiers, training situations, the *curricula* and educational paths have the effect of invoking apprenticeship as a tool for implementing professionalization. Through an

Encyclopedia of Professionalization,
coordinated by Didier DEMAZIÈRE and Richard WITTORSKI.

analysis of these different "wordings" of the concept of apprenticeship, the author's goal is to reveal the trends in the evolution of managerial and engineering discourses, regardless of whether they embody the engineering of training policies, social engineering, training, pedagogic or professionalization engineering. Based on an identification of the different contexts of use of apprenticeship, the text in section 6.2 proposes to show the interdependencies between different discourses emanating from work organizations and training organizations. This second section will end by questioning the added value expected from apprenticeship in terms of human resources management, (national and branch) training policies and the initial and ongoing training offer.

6.1. Contexts and conditions for the emergence of apprenticeship programs in education and in training

6.1.1. *Apprenticeship through the prism of everyday life*

Apprenticeship is a concept with such broad ambitions that it sometimes struggles to fit into daily practices. What forms does apprenticeship take when it enters into the thoughts and actions of training designers or managers? Regardless of its form, each apprenticeship program is structured around two poles: the school and the company. In other words, apprenticeship is often a constituent doctrine of professional training, whether it concerns the training of gardeners, doctors or bakers. Both in Quebec and Germany, apprenticeship constitutes training engineering at the service of the development of professional training and, more generally, of professionalization pathways in the different educational systems (high school and university). In France, apprenticeship is still specific to contractual arrangements serving policies to combat unemployment, or more recently, to reach the target of full employment. Remuneration for all (or part of) the professionalization path acts as a powerful incentive and exerts strong attractiveness among training students organized in cooperative programs. While apprenticeship constitutes one of the linguistic elements of the designers of professionalization paths, its organizational form boils down to associating, or even articulating, two distinct entities: a school or training center, on the one hand, and a company, on the other hand. We should mention that certain professional training courses and professions claim to be pioneers in the use of apprenticeship in the context of the training they offer: medical training, certain crafts and training in the agriculture sector. In this regard, the *Maisons familiales rurales* (MFR) have been communicating their particular engineering since 1934, which they summarize by the construct of an apprenticeship pedagogy.

6.1.2. *Apprenticeship and its ambitions to master spaces and times*

Apprenticeship training therefore embodies an undisputed marker of the technological and professional pathway, and even more so, in recent periods, of initial and ongoing professional training (Veillard 2017). Thinking about apprenticeship means inventing a strategic and organizational training method, whether it is presented as a scheme or as a path. First and foremost, evoking apprenticeship is imagining, in an ample sense, a process which organizes, structures, adapts and arranges a reality. By proposing to understand the concept of scheme, Aussel (2013) considers that it expresses an environment and accompanying situations. The notion of scheme evokes a space bordered by an environment and situations. We could also admit that the scheme expresses one or more temporalities (Pineau 2000). If we consider that, like the career path, the scheme is interdependent on apprenticeship, the organization's purpose, the development or adaptation proposed by apprenticeship means that it invites us to act on certain environments (situations which are already present); in other words, on spaces and times. Apprenticeship impacts space and time by proposing to reorganize them, structure them, rearrange them and adapt them.

The notion of career path is part of the heritage of the broader concept of career. In this view, the relationship of individuals to employment and work (Durand 2008) was assimilated to a long-term project, tied to the temporality of life's destiny. For its part, the notion of apprenticeship expresses a conception of time and social times. When apprenticeship is invoked (either as a scheme or as a career path), it expresses a distinct conception of spaces and a specific reading of temporalities. The articulation between apprenticeship, on the one hand, and the concepts of scheme and career path, on the other hand, stresses and characterizes a process of change with the goal of modifying and transforming a reality.

6.1.3. *Apprenticeship, environments and situations*

For several years, and all the more with the rise of professional didactics, the concept of situation, used in professional training, has attempted to bring together the double idea of time and space. The situation specifies the different moments or periods which structure the chronology of educational paths or professionalization paths. The situation characterizes the territory where the various potential training activities planned in the curriculum are envisioned and implemented. The situation mainly evokes the organization of the different associations and articulations between the school and the company, between an educational or training situation and a professional or work situation. For trainers or tutors, as for those responsible for training systems, the challenge is to orchestrate and plan a professionalization path by distributing the various situations in an argued and justified manner: those with explicitly training purposes and those exclusively considered as work situations. This organizational reading aiming at arranging the space-times of the school and those of the company continues to be the preferred entry point for defining apprenticeship. The situation is the key concept for apprenticeship primarily considered as a scheme that needs to be organized and managed.

Situations with an explicitly formative purpose can be training situations, potentially formative work situations, and internship situations (alternatively considered as work situations and/or training situations). In training, social and private situations are most often ignored. In the wake of reflections on apprenticeship, a new type of situation has recently emerged: the integration situation (Pentecouteau 2012). This involves planning situations explicitly aimed at integrating knowledge from internship experiences into training (Baartman and de Bruijn 2011). These integration situations are sometimes called integration workshops (Seezink et al. 2009). They use specific schemes or methods such as case studies, practice analysis seminars, accounts of practices, portfolio constructions, etc. The reference to the work of Varela (1988) and his theory of enaction (Durand 2008; Masciotra et al. 2008), invites us to reflect upon the links between cognition and situation, and leads to strengthening the assessment of situations from a cognitive learning perspective. Situations gradually become the *alpha* and *omega* of the training *curricula* and lead to the defense of situational engineering (Jonnaert et al. 2020, pp. 95–120), based in part on the legacy of work on professional didactic engineering (Mayen et al. 2017, pp. 467–482). Situations also bring out the dialogical and dialectical relationships between the different entities and the various protagonists involved in the scenography of apprenticeship (Bachelard 1957/2020).

6.1.4. *Three major philosophical debates echo three major questions posed to apprenticeship programs*

6.1.4.1. *The sensible world and the intelligible world*

First of all, being interested in apprenticeship involves seeking to extend the debate between the intelligible world and the sensible world (Jouët-Pastré 2013). This raises the question as to the relationship between a training project and a work project. It means agreeing to confront two situations whose respective logic differs. The training situation is thought of as an opportunity to give meaning to what we do, or what we plan to do. The work situation is the expression of doing in action. These two poles are organized as interdependent on one another.

The age-old opposition between the sensible world and the intelligible world impacts and even influences the conceptions of different training and work situations. The work situation is thought of as a situation involving direct and sensitive contact with reality. The training situation, for its part, is perceived as a situation capable of giving meaning to work situations (Barbier et al. 1996); in other words, of making them intelligible. As of late, professional didactics – or more generally, the theories of analysis of work activity (Thievenaz in this publication) – defend an intelligence of situations, while seeking to introduce doors to the intelligible world into the very heart of the work situation.

Access to the intelligible world expresses the difficulties of making the realities of the sensible world intelligible. Compared to the context of apprenticeship, the very exercise of a human activity obstructs the project of making such activity intelligible. This leads to considering a revisited apprenticeship in which intelligibility situations favoring greater articulation between work situations and training situations could be implemented.

6.1.4.2. *Poiesis and praxis*

The second philosophical debate echoing a problem raised by apprenticeship is that opposing *poiesis* versus *praxis*. This opposition stresses a strong trend, considering that apprenticeship is an organization aiming to articulate theory and practice. Theory and practice are not thought of as two distinct forms of knowledge, but rather as having two different purposes. From this perspective, the theory, as *poiesis*, aims at the finality of human action. It is at the service of an anticipated construction, but also analyzed a posteriori of a projected reality. *Poiesis* invites us to distinguish the means from the ends of production. *Praxis*, on the other hand, presents itself as a practical action with no other purpose than production itself. It refuses to submit to any transcendence. This distinction between *poiesis* and *praxis* permeates the common distinction made between theory and practice. Theory is subject to all constructions, all anticipations and all projections. It is one of the conditions of access to the intelligible world. It puts lived experiences into words (Barbier and Thievenaz 2013; Breton 2020). Practice, for its part, tends to remain confined to human activity, responding at the *here and now* of human needs. Fabre (1994) associates work with *praxis* and the oeuvre with *poiesis*. For Aristotle, in the Nicomachean Ethics, *poiesis* is a surpassing of action. It carries a creative intention within itself. *Poiesis* constitutes a "distance from the acts of production mobilizing the industrious body and its multi-determined actions" (Schwartz in Baudouin and Friedrich 2001, p. 71). The theories of action (Baudoin and Friedrich 2001) are part of the legacy of the thesis of a *praxis* driven by the project of responding effectively to the questions and even obstacles of daily human life. The theories for analyzing work activity embrace this epistemology (Champy-Remoussenard 2005, pp. 9–50; see also Chapter 9).

In order to overcome the dualism *poiesis/praxis* and its modern theory/practice variation, Dewey and other thinkers of pragmatism invite us to deconstruct all the dualisms that permeate our conceptions of education and training (Cometti 2010). Pragmatism defends the idea of practice without an opposite (Fabre 2015). Pragmatism also imposes a conception of the social and the societal inscribed in the realities of human existence. In the spirit of pragmatism, the only activity which counts is the one enabling the realization of the project of the person who drives it. This activity is not specified with regard to its practical dimension or its theoretical grounds. The purpose of the activity is to serve the economic and social development of societies. In other words, pragmatism as a post-modern philosophy shakes up our Hellenic philosophical heritage, which had led us so far to thinking about education and training (both from an ideological and political viewpoint and from an organizational and didactic-pedagogical viewpoint) as an opposition mode: education/training, general training/professional training, theoretical knowledge/practical knowledge, academic training versus technical training, intellectual versus manual, scholars/professionals, etc. Apprenticeship as a political, organizational and educational configuration crystallizes these different forms of dualism and often attempts to address them in vain. Pragmatism invites us to go beyond them.

6.1.4.3. *Adult training: its forms and its ends*

The third source at the origin of an interest in apprenticeship embraces the very grounds for adult education and/or training. In the following pages, we will offer an analytical, interpretative and comprehensive reading of adult education/training in order to uncover the different uses of the concept of apprenticeship.

We will notice that as far as apprenticeship is concerned, political discourses reflect ideologies and utopias anchored in different sociohistorical and sociopolitical contexts. Sometimes, apprenticeship is a term hiding behind other concepts or constructs, such as professional training (for example, Prost 2001). At other times, apprenticeship is the substitute for other words. Thus, when we read the term apprenticeship in speeches, other terms appear, such as company, professional training or even professionalization, in these days. Zealous speeches promoting the development of apprenticeship can be analyzed as pleas in favor of other arguments. To extol the merits of apprenticeship amounts to highlighting the company's eminently formative role. Proposing to develop apprenticeship is to magnify vocational training, often considered as a training path reserved for students with the most difficulties, particularly in France. Urging education and training professionals to include apprenticeship as a priority in their strategic planning is consistent with the idea of transforming educational paths into professionalization paths. Apprenticeship alone has brought together the different purposes assigned to adult training since the 1970s.

Apprenticeship programs are aimed at the professional training of adults (Sonntag 2019), the professionalization of métiers and educational paths (Fourdrignier 2007), qualification (Cahuc and Ferracci 2004; Solar and Tremblay 2008), certification (Kittel 2014), the recognition of experience (Zaouani-Denoux 2011), the analysis of activities and situational learning (Maubant 2013; Lecefel 2020), professional identity (Cohen-Scali 2000), professional development (Merhan 2014), professional transition (Perez-Roux 2013, pp. 325–328) and professional transmission (Roger 2015). It echoes the publications treating the thesis defended by Billett (Chapter 8): *Learning in the workplace*. Apprenticeship carries all these ambitions within itself. However, in recent years, it has tended to avoid the question of why, while preferring to answer the question of how.

For several years, research on apprenticeship has mainly focused on its organizational dimension (Clénet 2002; Landry 2002). How can we improve apprenticeship, understood as a scheme at the service of greater synergy between training and employment (Landry and Mazalon 1995)? How can we train the different actors of apprenticeship to make a scheme more effective, to make it increasingly demanded by policies (and this across all countries), and to deploy it in all training contexts welcoming adult audiences (Draper 2001, pp. 14–30; Zaouani-Denoux and Mazalon 2019)?

Even when it comes to studying apprenticeship pedagogy(ies), the analyses focuses on training engineering or pedagogical engineering questions (Fabre 1994, 2014, pp. 501–512). In recent times, it is the objects relating to the in-between that seem to focus research on apprenticeship (Astier 2007). These in-between periods are the internship periods or all the situations favoring synergies between training and work situations ultimately at the service of business development (Crocquey 1995; Deguerry and Trocquemé 2018). Research, and especially, research adhering to a utilitarian orientation to reveal best practices, is looking for tools, guides or instruments so as to better articulate all the situations constituting a professionalization path, in particular, internship situations.

6.1.5. *The ambitions of apprenticeship*

In concrete terms, the question hardly addressed until now by research on apprenticeship is that of why. Why use apprenticeship for adults? What are the benefits of apprenticeship for trained audiences? Can we justify the use of apprenticeship without first identifying its characteristics, its different designs and its various purposes? Are we able to characterize an educational configuration specific to apprenticeship? More recent research aims to demonstrate the rationale for apprenticeship in the context of situational learning at work (Maubant 2013; Jorro et al. 2017). Still others propose to question apprenticeship through the prism of experiences (Mayen and Maheux 2003) and temporalities (Maubant 2018).

6.1.5.1. *Experiences, situations and apprenticeship*

One of the keys to understanding the success of apprenticeship, if we consider it as the scenography for professional learning, lies in the identification and taking into account of the different experiences encountered by trainees, and to a lesser extent, by trainers. It should be noted that research on experience (Thievenaz 2019) as on transformative learning is rooted in American pragmatism, where the publications by Dewey (2018, 2022) and Kolb (1984) intersect those by Argyris and Schön (2001), advocating the idea that experiences must be analyzed in the light of the meaning attributed to them by human beings. In this spirit, Mezirow (1991) proposed using the meanings attributed to different lived experiences to draw up new forms of interpretation to orient future action. The production of meaning from experiences is the result of social interactions. It is subject to environments. For Dewey (2022), learning is necessarily experiential. It is also intimately related to the subject's situational action. The concept of situation then takes on its full meaning. The situation constitutes the framework within which lived experiences are interpreted. It is a catalyst for the different analytical tools that the subject summons to act in a situation. In the context of an apprenticeship program, the analysis and making sense of experiences are nourished by different contexts and experiential situations.

The process of interpreting lived experiences is also subject to the different temporalities identified and perceived by the subject. The various players in apprenticeship are encouraged to exploit any situation experienced by trainees for at least three reasons. The situation reveals lived experiences, and provides a framework for transformative learning. It also highlights the relationship to different temporalities. It suggests going beyond an overly linear reading of temporalities, while evoking the thesis of the subject's comings and goings throughout different lived times. "The relationships between past, present and future in learning constantly intersect in a complex way, inviting a more dialectical vision of the temporal dimension of learning" (Bourgeois 2013, p. 33). To support this argument, we can consider that apprenticeship adds a level of difficulty for managing the interactions between situations, learning, experiences and temporalities. This is because it invites both the trainee and the trainer to take into consideration several experienced situations having a different status, nature and form: social situations, training situations, work situations and internship situations. When exploring the concept of situation, it is necessary to uncover the double problem of experience and temporalitics.

Situation and experience are two concepts very present in Dewey's thought. In this regard, the two notions are representative of pragmatism defined as a philosophy of action. For Dewey, an experience is the accomplishment of a process constructed from an encounter with an undetermined, senseless situation, which gradually turns into a sensible situation, determined by the survey's logic.

For its part, experience expresses the general flow of life (Ogien 1999, pp. 69–75). Dewey does not differentiate understanding from experience. The thinking subject and experience are one. As for the situation, Dewey considers that it only exists through the encounter between the individual and the environment. In this perspective, the situation is directly articulated with the subject, through experiences. For Dewey, experience and situation are intimately related, linked dialogically and dialectically (Maubant and Roger 2017). These are two interdependent realities. Thus, the situation is imposed onto the subject. Indeed, the individual cannot refuse to confront a situation (Quéré 1997, p. 165). Experience is conditioned by it; however, this presupposes tackling a survey. This is then thought of as a process transposing the situation into an experience. Experience endows the situation with sense. It offers significations to it. Experience and situation constitute the framework of an educational proposal formulated by Dewey, aimed at articulating (or even uniting) the individual and the environment. The context in which the subject thinks and acts is considered by Dewey as a factor of emancipation. This proposal expresses the project of American pragmatism assigning to education the function of promoting encounters between the individual and its environment. Even being a philosophy of action, pragmatism defends two dualisms: individual-third party, on the one hand, and situation and experience, on the other hand. It gives a preponderant place to temporalities as markers and as analyzers of an experience and a situation.

6.1.5.2. *Temporalities, experiences and apprenticeship*

Another key to the success of apprenticeship, if it is given a professional learning aim, would involve the adequate management of temporalities (Dubar 2014). The thesis of *transformative learning theory* defended by Mézirow (1991), and developed by Alhadeff-Jones (2019, pp. 93–109) or Biasin (2018), allows us to focus the analysis on the necessary acknowledgment of temporalities by the different actors of apprenticeship, as an aid to learning and as an inducer of transformative learning. According to the authors who defend the transformative learning theory, temporalities "characterize transformation processes, likely to be promoted or supported in adult training" (Alhadeff-Jones 2018, p. 43). They interact with learning processes in four different ways. Alhadeff-Jones (2017) highlights the institutional and organizational temporal constraints having an impact on learning. Individual rhythms must also be taken into account in the analysis and accompaniment of learning. If we favor the thesis of transformative learning, once again, among those trained, rhythms affect the stages of their development and the markers of their transformation. Finally, the relationship with temporalities and the different experiences of perceived and lived times constitute significant variables in the conduct and accompaniment of transformative learning. We thus notice that a certain reading of apprenticeship, conceived as a set of interactions between temporalities, transformative learning and experiences opens up new questions as to links or even tensions between, on the one hand, times and rhythms and between times and experiences, on the other hand.

For Dewey (2018, p. 103), education is a process "reconstructing or reorganizing experience which adds to the signification of such experience and increases the capacity to direct the trajectory of subsequent experiences". It is worth emphasizing that temporalities are at the heart of Dewey's work.

Indeed, the interactions or transactions between situation and experience, desired by Dewey, are part of an adaptation project of the individual to the environment. From this perspective, this adaptation–transaction is done as the survey unfolds. As experience designates not only the adaptation–transaction process, but also the result of this process, Dewey gives two dimensions to it: one active and the other passive. The active dimension of experience expresses the role of ideas in the construction and exploitation of experience. The passive dimension refers to the sensitive world to the stage of the subject's first contact with the environment.

For Dewey, acting in and through the experience is a series of comings and goings in consciousness, in virtue of what it enables for anticipation and exploration and what it suggests as an explanation for lived experiences. These comings and goings are part of a discontinuous and nonlinear approach to time. They bring together in the same movement the time thought and the time lived; therefore, different temporalities. This introspective work on the experience favors the awareness of a double time for the subject, these times functioning asynchronously. The survey is based on a chronology. It calls on different times which are reunited for the same purpose: the production of knowledge resulting from an intellectual activity questioning experiences.

Apprenticeship reveals and crystallizes our different relationships with time and temporalities. The analysis of these relationships with the problem of time and times reveals different constituent characteristics of temporalities: linear and continuous time/disruptive and discontinuous time; time thought/conceived/lived/perceived; constrained time/chosen time (Aubert 2003); analysis time/action time; and time for self-reflection/time for the project (Boutinet 2012, pp. 361–376). In the context of an apprenticeship program, the translation of the different relationships to time and times is organized according to a rationality close to the one structuring and giving meaning to social times. However, this translation adds a managerial and bureaucratic conception of training times. The structuring of training *curricula*, the distribution of the different training periods (at the training center and at the company), the place and share of the periods reserved for "situational learning", in particular the internships, the evaluative, formative and certification moments, constitute the markers defining the framework and constrained nature of time in the educational path, or even in the professionalization path.

Following an analysis of the foundations of the discourse of institutions and organizations, this first section made it possible to identify the ideologies and utopias at the origin of the invocation of the concept of apprenticeship.

We will retain the following points from this first section. Apprenticeship is first of all a doctrine, on the boundary between *doxa* and the slogan. It functions as a "hook" for professional training systems. Apprenticeship also expresses an ambition, that of embracing the dimensions of space and time in an overall perspective. In other words, when a professional training scheme is publicized as apprenticeship, its commercial and pedagogic value is based on the affirmation of the presence of different training premises and training times. This valorization in itself constitutes the singularity of a training engineering and a pedagogic engineering. Apprenticeship is also an opportunity to revisit certain concepts: situation and experience. Thus, the work situation emerges as "the" situation around which apprenticeship can and must be organized. This work situation is expected to unfold its double capacity as a training situation and as a professional learning situation. As for the notion of experience, apprenticeship takes it into account by considering that it is a bet, in the sense that this singular engineering integrates and includes the different human experiences, even favors the dialogue between them, a dialogue which can lead to creating the conditions for professional learning. All of these elements highlighted in the first section of the chapter encourage us to emphasize to what extent the engineering of apprenticeship carries within it an educational engineering centered on the act of learning. Learning on an apprenticeship basis becomes the ultimate ambition of the designers of professional training systems and the various actors responsible for implementing them.

Section 6.2 will present the different reasons at the origin of the widely defended thesis of the presence of an apprenticeship pedagogy in various conceptions, discourses and practices. When reading research and publications on apprenticeship programs, apprenticeship pedagogy is extensively praised for its virtues. This specific pedagogy is expected to be a solution to four main problems, which are all interrelated: the training–employment relationship, the school–company relationship, the school–society relationship and the theory–practice relationship. These four problems, associated with their epistemological foundations and variations, have in common a relationship that is sometimes thought of as antagonistic and sometimes as complementary. They are the sources of questions. They crystallize tensions. They generate strong expectations. These apprenticeship pedagogies share the project as a common denominator, in short, as the promise of reconciling, articulating, bringing together and uniting two entities representing different statuses, functions, aims and organizational logics. Above all, they have an ambition, what Reboul (1992) calls the value of learning.

6.2. The apprenticeship pedagogy(ies): foundations, conceptions and uses

6.2.1. *From the learning pedagogies to the apprenticeship pedagogies*

Early on, adult training or education (Ministère de l'Éducation du Québec 2007; Keogh 2009) became interested in learning questions (Billett 1994; Billett and Pavlova 2005; Bourgeois and Nizet 2005; Groen and Kawalikak 2013, pp. 29–38), the field of transformative learning (Illeris 2004; Taylor 2007) and more recently, professional learning questions (CEDEFOP 2012; Durand and Bourgeois 2012; Troger and Ruano-Borbalan 2017, pp. 74–89) and situational learning at work (Marsick and Watkins 1990; Hodkinson and Hodkinson 2004, pp. 259–275; Tynjälä and Häkkinen 2005; Bourgeois and Mornata 2012, pp. 33–51; Fotius and Pagès 2014, pp. 89–97; Billett and Filliettaz 2015; Horcik and Durand 2015). Let us emphasize that research and scientific publications discussing the links between learning, professional learning and apprenticeship programs only rarely treat apprenticeship as a prism or analyzer of learning, nor the act of learning as a revealer and analyzer of apprenticeship. On the links between learning and apprenticeship, we can identify the work of Geay (1990), Candau and Kittel (2001, pp. 97–112), Bourgeois and Nizet (2005) and Bourgeois and Enlart (2014). Often, researchers study other objects to get closer, quietly, to the study of learning situations and processes in the context of professional training. Let us mention in particular the didactic and pedagogic perspectives proposing to treat learning in the context of an apprenticeship program from the viewpoint of the didactician and the pedagogue (Bourgeon 1979; Malglaive and Weber 1983; Geay and Sallabery 1999; Champy-Remoussenard 2005, pp. 319–336). Other research objects are also studied, but the frontal treatment of learning questions is not the chosen path. These are in particular the following objects: internships studied as a period or situation of particular training (Glaymann 2014; Zaouani-Denoux 2014), work and its activities (Veillard 2009, pp. 125–156), work and its analysis (Lhez et al. 2001), support for trainees or interns (Denoyel 2007, pp. 149–160; Leclercq et al. 2014), training systems or schemes (Zaouani-Denoux and Mazalon 2019), the training practices of trainers including tutors or mentors (Lebel et al. 2019), the training of trainers and business tutors (Lechaux 1992; Faingold 1996) and knowledge transfer (Maubant 1998; Pereira 2018) without forgetting, of course, training engineering (Leclercq 2003; Maubant 2020, pp. 11–13).

Vocational training contexts are often identified as intrinsically apprenticeship contexts. They are also used for treating situational learning at work (Lainé and Mayen 2019). Therefore, professional training and apprenticeship seem to constitute the same educational idea and, to say the least, a vast area of research and experimentation, especially around the notions of vocational education (Lyons et al. 1991) and vocational training (Wollschlager et al. 2004; Wheelahan 2021). They are both divided into training programs, educational paths and specific schemes aimed at three goals: professional learning, professional development and professional integration. Even the technological and professional path is today thought of and mandated as an apprenticeship path (Tanguy 2000). This is all the more important given the fact that various legislative and regulatory measures have recently opened up and strengthened the possibility of creating training within secondary establishments, where the trainee is not under an academic status but under an apprenticeship contract or professionalization contract[1] (Bertrand 2000, pp. 54–70; Arrighi and Mora 2010).

From these different modalities, researchers produce reflections on apprenticeship, and sometimes even proposals for educational reengineering (Guillaumin 2005). It is possible to identify from which angles scientific research proposes to highlight the different added values of apprenticeship. Publications insist on the fact that apprenticeship programs promote professional learning and learning altogether (Lerbet 1996; Savoyant 2010; Buchs and Bourgeois 2017, pp. 329–345; Faulx and Danse 2021; Bourgeois and Nizet 2005; Durand and Bourgeois 2012).

1 An apprenticeship contract and a professionalization contract are two types of employment contracts in France mainly aimed at young adults in training. In these contract schemes, the apprenticeship student is considered as an employee. As such, the student receives a remuneration paid by an employer. The latter benefits from reductions in employer social charges, which constitutes an incentive for the development of such contracts.

In this perspective, demonstrations aim to attest to the impacts of apprenticeship on cognitive processes (Savoyant 2010)

However, other publications also highlight how apprenticeship contributes to questioning, analyzing and producing new scientific knowledge on other objects: the professional identity of trainees (Kaddouri 2019, pp. 66–69; Cohen-Scali 2000), the training project and/or the professional project (Pineau 1983; Kaddouri 2002), teaching or training practices (Geay and Sallaberry 1999; Vanhulle et al. 2007), the didactic or pedagogical act (Guillaumin 2016), temporalities (Boutinet 2005; Roquet et al. 2013; Zaid and Lebeaume 2015; Perrinel and Maubant 2018, pp. 217–230) and training design (Maubant 2021). Apprenticeship training is a pretext for taking over the study of other objects, at the risk of sucking them into an educational utopia whose research struggles to identify the purposes, configurations and characteristics.

6.2.2. *Images, words and forms of apprenticeship pedagogies*

Today, resorting to Dewey's theses is partly explained by the desire to invoke a theory of learning articulated with real situations; in other words, with social situations and by the desire to place learning in a perspective of cultural and democratic development.

We could consider that, like pedagogy, apprenticeship pedagogy today finds itself torn between administrative progressivism and educational conservatism. In fact, the apprenticeship pedagogy is called upon to reorganize the educational path. From this perspective, apprenticeship pedagogy is often presented as a novelty, if not a real educational innovation. This makes it possible to invite the different partners to review and improve the unfolding of training by seeking a better match between the different training situations. This approach aims to define a regulatory framework specifying the prerogatives and responsibilities of each person in the proper functioning of training. In point of fact, this new organization of training struggles to go beyond the strict re-engineering. It fails to transform pedagogic practices, which remain locked in a traditional conception of the act of training. In other words, by getting stuck in a vain quest for an ideal method, apprenticeship pedagogy only extends the traditional pedagogy haunted by the reproduction of certain of its artifacts, notably the use of methods having proven their effectiveness. In this view, are certain tools or certain approaches used by apprenticeship pedagogy the new avatars of traditional pedagogy? Would not apprenticeship in its educational variation offer a veneer of modernity (Fabre 1994, 2014) to traditional pedagogy? Let us list some signs of this appearance: a teacher-centered logic, a valorization of learning by *mimesis* and the establishment of initiatory accompaniment. These signs reinforce the idea that the different pedagogic configurations of apprenticeship are part and parcel of the teaching process, even if they claim to be part of the training process (as in the administrative progressivism nourished by a *credo* and of *doxa* borrowing from psychology and the New Education) (Houssaye 2000). Apprenticeship pedagogies often express the hope of promoting the educational relationship at the very heart of the professionalization process. This hope finds its foundations in the histories of New Education and Popular Education. However, the various research studies that have analyzed the training practices of the different apprenticeship actors stress the extent to which such practices remain framed by a schooling conception of the act of training. Today, the theories of workplace learning (Rainbird et al. 2004) have a common denominator in the principles which echo those defended by the different variations of apprenticeship pedagogy: a focus on the individual learner, the consideration of cognitive sciences in the search for performance at work (Gollac and Volkoff 1996) and situational learning at work being taken for granted (Hager 2011, pp. 17–31). While the concept of competence has long been one of the keys to workplace learning theories, recent publications appeal to the concept of practices (and its different variations) meaning making practices, confessional practices, discursive practices, vocational practices, homemaking practices to highlight the vast heterogeneity situational learning at work (Griffiths and Guile 2001). The purpose of skills development remains one of the ambitions of thinkers encouraging workplace learning (Illeris 2011, pp. 32–45; Inffo Formation 2018).

6.2.3. *From learning models to apprenticeship pedagogic configurations*

Learning pedagogies revive an ancient question: the relationship between school and life (Houssaye 1987). At the same time, these pedagogies of the learning act invite us to question the different ways of learning. Analyzing apprenticeship in its different educational configurations in the light of learning pedagogies allows us to understand two of its major challenges: (1) the ideological, political and strategic discussion between the school and other social and human organizations; (2) the didactic, psychopedagogic and pedagogic discussion between different conceptions and professional learning models. Each of these discussions can still give rise to debates and even controversies.

In this sense, the learning model favored by the traditional school is based on the following triptych: observation; contemplation; impregnation. The educational *doxa* of observation–contemplation–impregnation will be supplemented by that of observation–imitation–realization. These *doxa* reveal a conception of learning (and more generally, a conception of professional learning), which will be fully displayed in the engineering of apprenticeship. *Mimesis* (Wulf 2014) embodies this double expectation between the educator and the student: efficiency in action and speed in learning.

6.2.4. *Analytical theories of the work activity: new perspectives for thinking about an apprenticeship pedagogy*

The success, popularity and interest these theories elicit, and the research they help to initiate, stress the extent to which a professional learning engineering seems to replace a professional didactic engineering (Pastré and Vergnaud 2011, pp. 401–421), a professionalization engineering (Cristol 2006) and a situational engineering (Duclos 2021), the latter having replaced skills engineering in recent years (Le Boterf 1999). We observe that professional learning engineering heralds an apprenticeship pedagogy engineering, which surpasses apprenticeship engineering. To serve this professional learning project and the ambition to think about an apprenticeship pedagogy, the analytical theories of the activity (inspired in work analysis) are invited into the discussion. They also invoke the concept of situation.

6.2.4.1. *From workplace learning to situational learning*

We notice to what extent most of the analytical theories of the activity study the activity's explicitation; in other words, a form of verbalization of the practice, to try to reveal this practical intelligence specific to each actor, but which can engage different invariants operating for the same task, regardless of the environment where it is carried out (Champy-Remoussenard 2005). By imposing a cognitive reading of the work activity, researchers gradually establish the question of learning according to the frameworks of work psychology (Clot 1999) and the psychology of learning (Olry 2019, pp. 105–121). The analysis of the situational activity becomes the opportunity to study the cognitive processes at work while carrying out such activity, from the initiation or discovery phase to the completion and success phase (Vergnaud 2011, pp. 275–292). Numerous North American works argue in favor of supporting the thesis of learning at work, or more precisely, defend the idea of learning in and through work (Billett 2001, 2002 and 2011; see Chapter 8; Engeström 2011, pp. 86–104). This leads to a proposal to reconsider training *curricula* according to this new learning paradigm, constitutive of the work situation (Billett 2006). The same goes for designing situations or work environments which promote learning (Billett 2004). In Europe, similar work is being developed by promoting the beneficial effects of learning on productive and learning organizations (Appelbaum and Gallagher 2000; Fuselier and Maroy 2002; Fischer et al. 2004). Situational learning is the subject of significant research, as evidenced by the work of Enlart, all the more because the dialogues suggested between the situation and learning make it possible to give shape to the professionalization process (chapter in this book).

6.2.4.2. *Interdependence between the situation and the activity*

By proposing to capture the work activity, research underlying the analytical theories of the activity will state that (1) the activity is a situated action: the worker confronts the activity with reality, with the situation, and with the resistance and constraints emanating from the environment; (2) the activity is nourished by the environment and the situation in which it takes place: the activity is regulated by the information constituting the situation. The individual engages not only practical intelligence but also strategic intelligence, because of which they will adapt the activity to the situation's contingencies (Pastré et al. 2006). Most often, this provides information on time, place, contexts as well as the activity's success criteria, set by labor, in tacit agreement (implicit or not) with the success criteria imposed (or not) by the contractor; (3) the activity is reflected upon and carried out in the interactions with others: contractors, supervisors, partners and peers are involved in carrying out such activity. In a number of situations, the worker is led to act in relation to (or even in collaboration with) others. They must think and organize the activity according to the others' activity; (4) the activity is organized according to social standards which correspond to the perception each person may have regarding the quality of an achievement (product or service) and the "beautiful oeuvre" they must "create". It also has to comply with the standards tacitly or explicitly imposed by the work context (Leplat 1997). The training of trainers in apprenticeship pedagogy cannot do without training in the analysis of the work activity, which could possibly support the idea of situational learning, well beyond work situations (Harris and Chisholm 2011, pp. 373–384; Durand 2012).

6.3. Conclusion: apprenticeship, stage director of professionalization paths and critical friend of life paths

While the activity's analysis, the promotion of reflexivity and the understanding of situations seem inseparably linked – in particular if we give apprenticeship programs the strict goal of fostering learning and professional development – what can be said about apprenticeship for thinking a re-engineering of professional training? Ease would lead us to answer that apprenticeship simply depends on the training context, on what the apprenticeship student achieves and the different situations in which they act. Then, it would merely suffice to put together different environments to be able to declare that we are in the presence of an apprenticeship program. However, this answer is far from satisfying because our conception of apprenticeship depends on our way of establishing the dialectic between the activity's analysis, reflexivity and the situation. In other words, apprenticeship, the activity's analysis and reflexivity are intertwined in an interdependence relationship. One has an impact on the other; however, apprenticeship nonetheless seems to provide the *tempo* for this dialogical and dialectical relationship between the activity's analysis, reflexivity and the situation. It shows a direction. It is apprenticeship which assigns a role to the activity's analysis, a role to reflexivity and sense to the situation. Apprenticeship orchestrates and stages these relationships. It organizes them according to its ambition and its project: the training-employment link, skills development or even professional learning.

Let us dare to extend the metaphor. Like any stage director, apprenticeship chooses its characters. It decides who will have the main role, who will have the secondary roles. Apprenticeship chooses the plots. It organizes situations. It takes into account settings in place and time. It accepts flashbacks. It outlines possible endings. But if we think of apprenticeship as if it were a stage director, the main thing is missing; in other words, the script. The script carries a message. It indicates a possible interpretation. It expresses a conception of reality and the world. Also, if apprenticeship is a script, its discourse reveals an intention: that of professional learning being possible, even desirable, regardless of the cost. If the apprenticeship stages the different formative situations, if it inserts into these situations actions analyzing the activity, if it attributes to each actor (in the sociological sense of the term and in the sense of the performing arts) a function, a role, a posture (that of a reflective professional), if it organizes the action according to different places and temporalities, if apprenticeship is invited to the main character's life path, it must indicate the sense of the play, or at least, its message or moral teaching.

According to this perspective, the sense of apprenticeship involves not only learning and professional development. By indicating this end of the play – which is acted out in professional training between the trainer, the expert professional and the novice professional – everyone knows their script. Each actor will know how to play its role and accept its character. Each actor will accept the setting. Everyone will assume they are part of the story. Everyone will submit to the different temporalities imposed on them by the stage director. However, everyone will also be able to open up the field of possibilities to the author's intention, a practical, functionalist intention or, on the contrary, a transformative and emancipatory one. Apprenticeship temporarily freezes in present time, while recalling past times and allowing a glimpse of the plurality of infinities.

6.4. References

Alhadeff-Jones, M. (2017). *Time and the Rhythms of Emancipatory Education. Rethinking the Temporal Complexity of Self and Society*. Routledge, London.

Alhadeff-Jones, M. (2018). Pour une conception rythmique des apprentissages transformateurs. *Phronesis*, 7(3), 43–52.

Alhadeff-Jones, M. (2019). Time and the rhythms of transformative learning. In *European Perspectives on Transformation Theory*, Flemming, T., Kokkos, A., Finnegan F. (eds). Palgrave-Macmillan, New-York.

Appelbaum, S.H. and Gallagher, J. (2000). The competitive advantage of organisational learning. *Journal of Workplace Learning*, 12, 40–46.

Argyris, C. and Schön, D.A. (2001). *Apprentissage organisationnel. Théorie, méthode, pratique*. De Boeck Supérieur, Brussels.

Arrighi, J.-J. and Mora, V. (2010). Contrats de professionnalisation et d'apprentissage : des usages diversifiés. Esquisse d'une cartographie des formations en alternance. Net. Doc. No. 66. CEREQ [Online]. Available at: https://www.cereq.fr/sites/default/files/2018-09/32bfdfd83705514c16513efebea6cf2a.pdf.

Astier, P. (2007). Alternance construite, prescrite, vécue. *Éducation permanente*, 172(3), 61–72.

Aubert, N. (2003). *Le culte de l'urgence. La société malade du temps*. Flammarion, Paris.

Aussel, L. (2013). Évaluer les dispositifs : le cas d'un dispositif de formation de l'enseignement supérieur agricole. Unpublished PhD Thesis. Université de Toulouse Le Mirail.

Baartman, L.K.J. and de Bruijn, E. (2011). Integrating knowledge, skills and attitudes: Conceptualising learning processes towards vocational competence. *Educational Research Review*, 6(2), 125–134.

Bachelard, G. (1957/2020). *La poétique de l'espace*. Presses Universitaires de France, Paris.

Balas, S. (2020). Place des certifications professionnelles contemporaines dans l'orientation tout au long de la vie. In *L'orientation tout au long de la vie*, Soidet, I, Olry-Louis, I., Blanchard, S. (eds). L'Harmattan, Paris.

Barbier, J.-M. and Thievenaz, J. (2013). *Le travail de l'expérience*. L'Harmattan, Paris.

Barbier, J.-M., Berton, F., Boru, J.-J. (eds) (1996). *Situation de travail et formation*. L'Harmattan, Paris.

Baudouin, J. and Friedrich, J. (2001). *Théories de l'action et éducation*. De Boeck Supérieur, Brussels.

Bertrand, O. (2000). L'évolution de l'alternance et de l'apprentissage ici et ailleurs. In *Actes de la Conférence internationale sur la formation professionnelle et technique. Les relations entre la formation et le monde du travail : enjeux et pratiques*, Payeur, C. and Bernard, Y. (eds). Centrale des syndicats du Québec. Montreal.

Biasin, C. (2018). L'apprentissage transformateur : état des lieux et portée heuristique d'un construit en développement. *Phronesis*, 7(3), 1–4.

Billett, S. (1994). Situated learning: A workplace experience. *Australian Journal of Adult & Vocational Education*, 34, 112–130.

Billett, S. (2001). Learning through work: Workplace affordances and individual engagement. *Journal of Workplace Learning*, 13(5), 209–214.

Billet, S. (2002). Workplace pedagogic practices: Co-participation and learning. *British Journal of Educational Studies*, 50(4), 457–481.

Billett, S. (2004). Workplace participatory practices. Conceptualising workplaces as learning environments. *The Journal of Workplace Learning*, 16(6), 312–324.

Billett, S. (2006). Constituting the workplace curriculum. *Journal of Curriculum Studies*, 38(1), 31–48.

Billett, S. (2011). *Vocational Education Purposes: Traditions and Prospects*. Springer, New York.

Billett, S. and Filliettaz, L. (2015). *Francophone Perspectives of Learning Through Work. Conceptions, Traditions and Practices*. Springer, New York.

Billett, S. and Pavlova, M. (2005). Learning through working life: Self and individuals' agentic action. *International Journal of Lifelong Education*, 24(3), 195–211.

Boudjaoui, M. and Leclercq, G. (2014). Revisiter le concept de dispositif pour comprendre l'alternance en formation. *Éducation et francophonie*, 42, 22–41.

Bourgeois, É. (2013). Entre aliénation et émancipation : les figures du sujet de l'expérience. *Travail et Apprentissages*, 12(2), 79–93.

Bourgeois, É. and Enlart, S. (2014). *Apprendre dans l'entreprise*. Presses Universitaires de France, Paris.

Bourgeois, É. and Mornata, C. (2012). Apprendre et transmettre le travail. In *Apprendre au travail*, Bourgeois, É. (ed.). Presses Universitaires de France, Paris.

Bourgeois, E. and Nizet, J. (2005). *Apprentissage et formation des adultes*. Presses Universitaires de France, Paris.

Bourgeon, G. (1979). *Socio-pédagogie de l'alternance*. UNMFREO, Tours.

Boutinet, J.-P. (2005). *Vers une société des agendas. Une mutation des temporalités*. Presses Universitaires de France, Paris.

Boutinet, J.-P. (2012). Des temporalités paradoxales aux temporalités tyranniques. In *Anthropologie du projet*, Boutinet, J.-P. (ed.). Presses Universitaires de France, Paris.

Breton, H. (2020). Narration du vécu et savoirs expérientiels. *Éducation permanente*, 222, 5–10.

Buchs, C. and Bourgeois, E. (2017). Conflits sociocognitifs et apprentissage. In *Traité des sciences et des techniques de la Formation*, 4th edition, Carré, P. and Caspar, P. (eds). Dunod, Paris.

Cahuc, P., Ferracci, M., Tirole, J., Wasmer, É. (2014). L'apprentissage au service de l'emploi. *Notes du conseil d'analyse économique*, 19, 1–12.

Candau, G. and Kittel, F. (2001). Situation pédagogique de l'alternance et construction du savoir de l'apprenant. In *Éducation et travail : divorce ou entente cordiale ?*, Perez, S. and Strobel O. (eds). L'Harmattan, Paris.

CEDEFOP – Centre européen pour le développement de la formation professionnelle (2012). *Apprendre en travaillant. Apprendre en contexte professionnel : exemples de réussite en Europe*. Office des publications de l'Union européenne, Brussels.

Champy-Remoussenard, P. (2005). Écritures sur l'activité professionnelle et pédagogie de l'alternance dans l'enseignement supérieur. In *L'alternance dans l'enseignement supérieur. Enjeux et perspectives*, Hahn, C., Besson, M., Colin, B., Geay A. (eds). L'Harmattan, Paris.

Clénet, J. (2002). *L'ingénierie des formations en alternance : pour comprendre, c'est-à-dire pour faire*. L'Harmattan, Paris.

Clot, Y. (1999). *La fonction psychologique du travail*. Presses Universitaires de France, Paris.

Cohen-Scalli, V. (2000). *Alternance et identité professionnelle*. Presses Universitaires de France, Paris.

Cometti, J.-P. (2010). *Qu'est-ce que le pragmatisme*. Gallimard, Paris.

Cristol, D. (2006). L'ingénierie de la professionnalisation. *Actualité de la formation permanente*, 200, 32–35.

Crocquey, E. (1995). Les capacités de formation des entreprises en 1995 : de l'alternance à la formation en situation de travail. *DARES Indicateurs*, 199, Paris.

Deguerry, N. and Trocquemé, C. (2018). L'action de formation en situation de travail réinterroge la définition de l'action de formation. *Inffo Formation*, 952, 2–3.

Denoyel, N. (2007). Réciprocité interlocutive et accompagnement dialogique. In *Penser l'accompagnement adulte*, Boutinet, J.-P., Denoyel, N., Pineau, G., Robin, J.-Y. (eds). Presses Universitaires de France, Paris.

Dewey, J. (2006). *Logique : la théorie de l'enquête*. Presses Universitaires de France, Paris [Original published 1938].

Dewey, J. (2022). *Démocratie et éducation : suivi de expérience et éducation*. Armand Colin, Paris [Original published 1916].

Draper, J.A. (2001). The metamorphoses of andragogy. In *Fundamentals of Adult Education. Issues and Practices for Lifelong Learning*, Poonwassie, D.H. and Poonwassie, A. (eds). Thompson Education Publishing, Toronto.

Dubar, C. (2014). Du temps aux temporalités : pour une conceptualisation multidisciplinaire. *Temporalités*, 20 [Online]. Available at: https://temporalites.revues.org/2942.

Duclos, L. (2021). L'action de formation en situation de travail (AFEST) : de l'expérimentation à la loi. Ministère du Travail et de l'emploi. Gouvernement de la France [Online]. Available at: https://www.oecd.org/els/emp/skills-and-work/adult-learning/France.pdf.

Durand, M. (2008). Un programme de recherche technologique en formation des adultes : une approche enactive de l'activité humaine et l'accompagnement de son apprentissage/développement. *Éducation & Didactique*, 3(2), 69–93.

Durand, M. (2012). L'alternance : une métaphore prometteuse d'innovation sociale et éducative. *Éducation permanente*, 193, 19–30.

Durand, M. and Bourgeois, E. (2012). *Apprendre au travail*. Presses Universitaires de France, Paris.

Engeström, Y. (2011). Activity, theory and learning at work. In *The Sage Handbook of Workplace Learning*, Malloch, M., Cairns, L., Evans, K., O'Connor, B.N. (eds). Sage, Thousand Oak.

Fabre, M. (1994). *Penser la formation*. Presses Universitaires de France, Paris.

Fabre, M. (2014). La pédagogie et les pédagogies. In *Traité des sciences et des pratiques de l'éducation,* Beillerot, J. (éd.). Dunod, Paris.

Fabre, M. (2015). *Éducation et humanisme. Lecture de John Dewey*. Vrin, Paris.

Faingold, N. (1996). Explicitation de pratiques tutorales et formation de formateurs. *Recherche et Formation*, 22, 81–97.

Faulx, D. and Danse, C. (2021). *Comment favoriser l'apprentissage et la formation des adultes ?* De Boeck, Brussels.

Fischer, M., Boreham, N., Nyhan, B. (eds) (2004). *European Perspectives on Learning at Work: The Acquisition of Work Process Knowledge.* Office for Official Publication of the European Communities, Luxembourg.

Fotius, P. and Pagès, S. (2014). Les pratiques d'apprentissage en situation de travail. In *Apprendre dans l'entreprise*, Enlart, S. and Bourgeois, É. (eds). Presses Universitaires de France, Paris.

Fourdrignier, M. (2007). Alternance et professionnalisation : le cas des métiers du social. *Marché et organisations*, 5, 79–100.

Fusulier, B. and Maroy, C. (2002). La formation en alternance en Belgique francophone : développements pratiques et théoriques. In *La formation en alternance : état des pratiques et des recherches*, Landry, C. (ed.). Presses de l'Université du Québec, Saine-Foy.

Geay, A. (1990). Stratégie d'apprentissage et stratégie d'enseignement en système de formation alternée. *Éducation permanente*, 106, 51–58.

Geay, A. and Sallaberry, J.-C. (1999). La didactique en alternance ou comment enseigner dans l'alternance ? *Revue française de pédagogie*, 128, 7–15.

Glaymann, D. (2014). Le stage dans l'enseignement supérieur, un dispositif riche de promesses difficiles à tenir. *Phronesis*, 3(1–2), 58–69.

Gollac, M. and Volkoff, S. (1996). *Citius, altius, fortius.* L'intensification du travail. *Actes de la recherche en sciences sociales*, 114, 54–67.

Griffiths, T. and Guile, D. (2001). Learning through work experience. *Journal of Education and Work*, 14(1), 113–131.

Groen, J. and Kawalilak, C. (2013). The tapestry of adult education research in Canada. In *Building on Critical Traditions: Adult Education and Learning in Canada*, Nesbit, T., Brigham, S., Gibb, T., Taber, N. (eds). Toronto Thompson Press, Toronto.

Guillaumin, C. (2005). Apprendre, ailleurs, autrement : construction et accompagnement d'une ingénierie pédagogique de l'alternance. *Éduquer*, 10, 117–127.

Guillaumin, C. (2016). Le sens de l'alternance. Alternative pédagogique, projet éthique, et perspective politique. *Phronesis*, 1(5), 28–37.

Hager, P. (2011). Theories of workplace learning. In *The Sage Handbook of Workplace Learning*, Malloch, M., Cairns, L., Evans, K., O'Connor, B.N. (eds). Sage, Thousand Oaks.

Harris, M. and Chisholm, C. (2011). Beyond the workplace: Learning in the lifeplace. In *The Sage Handbook of Workplace Learning*, Malloch, M., Cairns, L., Evans, K., O'Connor, B.N. (eds). Sage, Thousand Oaks.

Hodkinson, P. and Hodkinson, H. (2004). The complexities of workplace learning. Problems and dangers in trying to measure attainment. In *Workplace Learning in Context*, Rainburd, H., Fuller, A., Munro, A. (eds). Routledge Falmer, London.

Horcik, Z. and Durand, M. (2015). L'expérience mimétique dans l'apprentissage adulte : le cas des formations par simulation. *Schweizerische Zeitschrift für Bildungswissenschaften*, 37(1), 167–186.

Houssaye, J. (1987). *École et vie active : résister ou s'adapter.* Peter Lang, Bern.

Houssaye, J. (2000). *Le triangle pédagogique*, 3rd edition. Peter Lang, Bern.

Illeris, K. (2004). Transformative learning in the perspective of a comprehensive learning theory. *Journal of Transformative Education*, 2(2), 79–89.

Illeris, K. (2011). Workplaces and learning. In *The Sage Handbook of Workplace Learning*, Malloch, M., Cairns, L., Evans, K., O'Connor, B.N. (eds). Sage, Thousand Oaks.

Inffo Formation (2018). Les pratiques de développement des compétences en situation de travail. *Inffo Formation*, 948, 3.

Jonnaert, P., Depover, C., Malu, R. (2020). Une approche par les situations. In *Curriculum et situations. Un cadre méthodologique pour le développement de programmes éducatifs*, Jonnaert, P., Depover, C., Malu, R. (eds). De Boeck Supérieur, Bruxelles.

Jorro, A., De Ketele, J.-M., Mehran, F. (2017). *Les apprentissages professionnels accompagnés*. De Boeck, Brussels.

Jouët-Pastré, E. (2013). Que signifie voir l'intelligible dans les dialogues de Platon ? *Pallas*, 92, 217–224.

Kaddouri, M. (2019). Dynamiques identitaires. In *Vocabulaire des histoires de vie et de la recherche biographique*, Delory-Momberger, C. (ed.). Érès, Paris.

Kittel, I. (2014). Les référentiels, supports ou obstacles à la professionnalisation ? *Empan*, 95, 42–48.

Kolb, D. (1984). *Experiential Learning*. Prentice Hall, Hoboken.

Lainé, A. and Mayen, P. (eds) (2019). *Valoriser le potentiel d'apprentissage des expériences professionnelles : repères, démarches et outils pour accompagner l'apprenant en formation par alternance*. Éducagri éditions, Dijon.

Landry, C. (ed.) (2002). *La formation en alternance : état des pratiques et des recherches*. Presses de l'Université du Québec, Sainte-Foy.

Landry, C. and Mazalon, É. (1995). Évolution et tendances des relations formation-travail en formation professionnelle et technique au Québec : du flirt à la réconciliation. *Revue des sciences de l'éducation*, 21(4), 781–808.

Le Boterf, G. (1999). *L'ingénierie des compétences*. Éditions d'Organisation, Paris.

Lebel, C., Van Nieuwenhoven, C., Colognesi, S., Bélair, L. (2019). Introduction. Le travail d'accompagnement des formateurs de stagiaires dans l'enseignement supérieur : perspectives compréhensives et modalités de mise en place. *Phronesis*, 1–2(8), 1–4.

Lecefel, P. (2020). L'analyse de l'activité en situation professionnelle comme ressource dans la formation au CFA. *Travail et Apprentissages*, 20, 77–92.

Lechaux P. (1992). Former les tuteurs, les enjeux du professionnalisme pour les entreprises. *Actualité de la Formation Permanente*, 119, 62–69.

Leclercq, G. (2003). Quelques usages de l'activité d'ingénierie de formation. *Savoirs*, 2(2), 71–104.

Leclercq, G., Oudart, A.-C., Marois, T. (2014). L'accompagnabilité, une propriété des dispositifs de formation en alternance : des manières d'accompagner et de se faire accompagner. *Éducation et francophonie*, 42(1), 136–150.

Leplat, J. (1997). *Regards sur l'activité en situation de travail*. Presses Universitaires de France, Paris.

Lerbet, G. (1996). *L'école du dedans*. Hachette Éducation, Paris.

Lhez, P., Millet, D., Séguier, B. (2001). *Alternance et complexité en formation*. Éditions Seli Arslan, Paris.

Lyons, E., Randhawa, B., Paulson N. (1991). The development of vocational education in Canada. *Canadian Journal of Education*, 16(2), 137–150.

Malglaive, G. and Weber, A. (1983). École et entreprise : intérêt et limites de l'alternance en pédagogie. *Revue française de pédagogie*, 62, 51–64.

Marsick, V.J. and Watkins, K.E. (1990). *Informal and Incidental Learning in the Workplace*. Routledge, New York.

Masciotra, D., Roth, W.M., Morel, D. (2008). *Énaction. Apprendre et enseigner en situation*. De Boeck Université, Brussels.

Maubant, P. (1998). Transfert et pédagogie par alternance. *Éducations*, 15, 50–56.

Maubant, P. (2013). *Apprendre en situations : un analyseur de la professionnalisation dans les métiers adressées à autrui*. Presses de l'Université du Québec, Sainte-Foy.

Maubant, P. (2018). *Les temps heureux des apprentissages*. Champ social éditions, Nîmes.

Maubant, P. (2020). Préface. In *Formation par alternance. Au cœur de l'aventure humaine*, Clénet, J. (ed.). Champ social éditions, Nîmes.

Maubant, P. (2021). *L'archipel des temps de la formation : esquisse d'une ingénierie de l'alternance*. Champ social éditions, Nîmes.

Maubant, P. and Roger, L. (2017). Lecture pédagogique de l'alternance en éducation et en formation à la lumière des idées de John Dewey. *Questions Vives*, 21, 1–21.

Mayen, P. and Maheux, C. (2003). Expérience et formation. Note de synthèse. *Savoirs*, 1, 15–53.

Mayen, P., Orly, P., Pastré, P. (2017). L'ingénierie didactique professionnelle. In *Traité des sciences et des techniques de la Formation*, 4th edition, Carré, P. and Caspar, P. (eds). Dunod, Paris.

Merhan, F. (2014). Alternance en formation et développement professionnel : enjeux du portfolio. *Éducation et francophonie*, 42(1), 80–94.

Mezirow, J. (1991). *Transformative Dimensions in Adult Learning*. Jossey-Bass/Wiley, New York.

Ministère de l'éducation du Québec (2007). Une histoire de l'éducation des adultes [Online]. Available at: http://www.education.gouv.qc.ca/fileadmin/site_web/documents/dpse/educ_adulte_action_comm/41-7018.pdf.

Ogien, A. (1999). Émergence et contrainte : situation et expérience chez Dewey et Goffman. In *La logique des situations. Nouveaux regards sur l'écologie des situations*, de Fornel, M. and Quéré, L. Éditions de l'École des Hautes Études en Sciences Sociales, Paris.

Olry, P. (2019). Aleksei Leontiev et la théorie de l'activité. In *Psychologies pour la formation*, Carré, P. (ed.). Dunod, Paris.

Pastré, P. and Vergnaud, G. (2011). L'ingénierie didactique professionnelle. In *Traité des sciences et des techniques de la formation*, Carré, P. (ed.). Dunod, Paris.

Pastré, P., Mayen, P., Vergnaud, G. (2006). La didactique professionnelle. *Revue française de pédagogie*, 154, 145–198.

Pentecouteau, H. (2012). L'alternance dans une formation professionnelle universitaire. De l'idéal épistémologique aux contradictions pédagogiques. *Revue internationale de pédagogie de l'enseignement supérieur*, 28(1), 1–12.

Pereira, I. (2018). Les dimensions axiologiques du transfert des connaissances scientifiques dans la pratique professionnelle. Un exemple en formation des enseignants. *Spirale*, 61(1), 151–161.

Perez-Roux, M. (2013). Transition professionnelle. In *Dictionnaire des concepts de la professionnalisation*, Jorro, A. (ed.). De Boeck Supérieur, Brussels.

Perrinel, M. and Maubant, P. (2018). L'alternance comme révélateur de la fonction des temporalités dans la compréhension des situations d'apprentissage professionnel. In *Les temps heureux des apprentissages*, Maubant, P. (ed.). Champ social éditions, Nîmes.

Pineau, G. (2000). *Temporalités en formation. Vers de nouveaux synchroniseurs*. Anthropos, Paris.

Quéré, L. (1997). La situation toujours négligée ? *Réseaux*, 15(85), 163–192.

Rainbird, H., Fuller, A., Munro, A. (eds) (2004). *Workplace Learning in Context*. Routledge, London.

Reboul, O. (1992). *Les valeurs de l'éducation*. Presses Universitaires de France, Paris.

Remoussenard, P. (2005). Les théories d'analyse de l'activité : entre travail et formation. *Savoirs*, 2(8), 9–50.

Roger, L. (2015). *Mise en œuvre de la professionnalisation dans l'enseignement supérieur (alternance, e-learning, parcours)*. Champ social éditions, Nimes.

Roquet, P., Gonçalves, M.-J., Roger, L., Viana-Caetano, A.-P. (2013). *Temps, temporalité et complexité dans les activités éducatives et formatives*. L'Harmattan, Paris.

Savoyant, A. (2010). Une approche cognitive de l'alternance. *Travail et apprentissages*, 5(1), 119–126.

Seezink, A., Poell, R.F., Kirschner, P. (2009). Teachers' individual action theories about competence-based education: The value of the cognitive apprenticeship model. *Journal of Vocational Education & Training*, 61(2), 203–215.

Solar, C. and Tremblay, N. (2008). *Bilan des recherches en français au Canada sur l'apprentissage des adultes : 1997-2007*. Centre de savoir sur l'apprentissage, Ottawa.

Sonntag, M. (2019). La formation continue des adultes en France : des repères. *Les Sciences de l'éducation – Pour l'Ère nouvelle*, 52, 101–122.

Taylor, E. (2007). An update on transformative learning theory: A critical review of empirical research (1999-2005). *International Journal of Lifelong Education*, 26(2), 173–191.

Thievenaz, J. (2019). *Enquêter et apprendre au travail : approcher l'expérience avec John Dewey*. Raison et Passions, Dijon.

Troger, V. and Ruano-Borbalan, J.-C. (2017). Former au travail. Des corporations à l'enseignement technique. In *Histoire du système éducatif*. Presses Universitaires de France, Paris.

Tynjälä, P. and Häkkinen, P. (2005). E-learning at work: Theoretical underpinnings and pedagogical challenges. *The Journal of Workplace Learning*, 17(50-6), 318–336.

Vanhulle, S., Mehran, F., Ronveaux, C. (2007). *Alternances en formation*. De Boeck, Louvain-la-Neuve.

Varela, F. (1988). *Invitation aux sciences cognitives*. Seuil, Paris.

Veillard, L. (2009). Organiser les situations de travail pour l'apprentissage : approche didactique de la formation professionnelle par alternance. In *Travail et formation des adultes,* Durand, M. (ed.). Presses Universitaires de France, Paris.

Veillard, L. (2017). *La formation professionnelle initiale*. Presses Universitaires de Rennes, Rennes.

Vergnaud, G. (2011). Au fond de l'action, la conceptualisation. In *Savoirs théoriques et savoirs d'action,* Barbier, J.-M. (ed.). Presses Universitaires de France, Paris.

Wheelahan, L. (2021). Doctoral theses abstracts in vocational education. *Journal of Vocational Education & Training*, 73(4), 607–608.

Wollschlager, N. and Fries Guggenheim, É. (2004). A history of vocational education and training in Europe: From divergence to convergence. *European Journal: Vocational Training*, 32, 6–17.

Wulf, C. (2014). Mimésis et apprentissage culturel. *Le Télémaque*, 45, 123–136.

Zaid, A. and Lebeaume, J. (2015). *La formation d'ingénieurs en alternance. Rythmes et temporalités vécues*. Presses du Septentrion, Lille.

Zaouani-Denoux, S. (2011). La formation en alternance : effectuation, itinéraire et construction de soi. *Savoirs*, 26, 61–83.

Zaouani-Denoux, S. (2014). Apprendre en stage : situation de travail, interactions et participation. *Éducation et socialisation*, 35. doi: 10.4000/edso.642.

Zaouani-Denoux, S. and Mazalon, E. (2019). *La formation en alternance. Diversité des dispositifs, perspectives des usagers et complexité des approches*. L'Harmattan, Paris.

7

The Relationships Between Professionalization and the Work Situation: Professional Challenges and Social Relationships

Sandra ENLART

Apprenance Team, Paris Nanterre University, France

We did not have to wait until recently to become interested in on-the-job training. Throughout time, humanity has understood that it is at the heart of the activity that we learn our métier. Traditions tell us that experience, especially when we start working at a young age (Heinich 1993), is a necessary point of passage to access a profession, this being considered both as a social community and a field of expertise. In this sense, what is at stake in professionalization is also the reflection of a social, cultural and political organization, mirroring a certain society. Let us take the example of adult education movements. If there is one tradition which has been inherited from one revolution to another, from continent to continent, it is the affirmation of a right to education for the popular masses (Condorcet 1792). Thinkers as different as Dewey (1916), Vasconcelos (Borjon-Nieto and Lascurain-Fernandez 2016) or Freire (1968–1974), in completely different contexts, have each associated democracy with adult education. Conversely, life stories, considered as individual learning stories, are also the reflection of a sociopolitical context (Dominicé 1992). In this chapter, we will focus on professionalization models as analyzers of a sociopolitical organization. Do these models not characterize the

Encyclopedia of Professionalization,
coordinated by Didier DEMAZIÈRE and Richard WITTORSKI.

way in which we *must* be professionalized, meaning that we have to learn the trade before being professionally and socially recognized?[1] What is learned is both a profession and a position in the social order. Also, the analysis of political contexts, economic orientations and industrial relations makes it possible to understand the meaning of adult education and training policies, as was done by Ashton (2004). Although our analysis will be partly historical, it will not deny the coexistence of different models. We will discuss three visions of professionalization schemes in the light of the social relationships they reflect. The first insists on the fact that learning a profession also involves integrating a social order. The second one envisions training as an individual right, an approach mastered by an employee engaged in professionalization, free from any traditional order. Finally, a third approach considers professionalization as a process at the heart of the workplace and the activity, facilitated by an organization which is itself learning.

7.1. The social order at the center

This first model is characterized by the same idea at different periods and in varying contexts: the work situation is not only an opportunity to learn a métier, but also to internalize social relationships (hierarchy, power and submission).

7.1.1. *When to learn is to obey: compagnons, artisans, workers in history*

Before approaching contemporary organizations, we must recall the way in which a métier was learned in older times through *compagnonnage*. It was a long journey, during which almost total submission to the apprenticeship master was the rule. Even if it is somehow idealized today[2], the realities of life and work were particularly trying. Characterized by a strong, unchanging hierarchy, working conditions were extremely harsh: long hours, very low or no pay, challenging physical environment, routines, standardization of phases (Heinich 1993). Beyond the specific apprentice–master relationship, it is more fundamentally the place and role of corporations which organized professional relations in society. Since then, we have understood why the French Revolution worked so hard to deconstruct that social order[3]: the social organization seemingly had to be brought down so that the individual citizen could emerge: "there are no more corporations in the state; there is only the interest of each individual and the general interest" (Archives parlementaires, 1st series, Volume XXXI, pp. 617–619). This coincidence between

1 The answers are not identical, depending on where we are placed in the social hierarchy.

2 For example, in the way of presenting the *Compagnons du Devoir* in France.

3 In particular with the Le Chapelier law of June 14, 1791.

professionalization modes and the organization of power at all levels is just as visible in the life story of a *compagnon* in the 17th century as in the revolutionary desire to destroy such order.

At the beginning of industrialization, lack of autonomy was still the rule. In fact, we can barely speak of "training", so harsh were working conditions. It is necessary to reread Jules Simon's (1867) study on child labor to realize the degree of exploitation and violence that existed in the emerging industry. "In Normandy, the cosh was included under the working instruments of the métier". Dependence was total on both the father and the boss. Perrot (1996, p. 275) thoroughly describes the ambivalent relationships between employers and the family. "In all cases, in the confrontation between the working world and the employers, the family is a strategic issue. At the center of the conflict, young people are both protected and directed, supported and commanded by this ambivalent reality embodied by the family." Dependence on the father–apprenticeship master is all the more central as professionalization takes place over a very long time. In point of fact,

> the heritage of working families is the métier, or at least the wages... In the blast furnaces, children first work as servants, then around twelve years old, they become 'valets' (*goujats*) alongside the refiners; they remain without a place for a long time, without the certainty of having one, but the hierarchy and family precedence are not contested by anyone (Perrot 1996).

This transmission relationship structured society as a whole, making any rebellion impossible. Transmission and dependence went hand in hand, like a sort of barter between those who knew how to do it and those who had to wait for years, if necessary as a "newbie" (*arpete*), sleeping in the attic, mistreated, and without pay for 4–5 years. Of course, there is not much in common between the apprentice at a painter's workshop at the beginning of the 16th century in Italy and the apprentice worker in the 19th century English factory: only this shared vision of learning as a long-haul effort, inevitable, subject to a hierarchical social organization, and impossible to call into question. These features are the faithful reflection of the social and political organization that prevailed until the very beginning of the 20th century. Despite the impossibility of comparing such different periods and countries, the fact remains that continuity exists in power relationships, the social classes structure and the access to the profession. It is with systems such as the taskers that a rupture appears. These were workers "who dealt with the merchant and divided labor to distribute it among the other workers or the workshops" (Abdelnour and Méda 2019). The learning paths then disappeared, as workers hired for very short periods had to follow strict instructions to the letter on a task reduced to a minimum.

In the middle of the 19th century, this English or French sweating system isolated but also deskilled workers and broke the logic of trades and professionalization. It somehow prefigured Taylorism (Barraud de Lagerie and Sigalo Santos 2019).

7.1.2. *At the end of the war: when to learn is to apply*

Now let us enter our modern world. At the heart of the war, and then at its end, in the United States, it became necessary to produce quickly and in great amounts. It was essential to train as quickly as possible to master the right gesture, and only the right gesture. Learning by doing and TWI[4] would become the ideal model: effective learning limited to the strict needs of production. This learning perfectly corresponded to the working model at the Fordist factory: designed by the methods engineer and imposed on the worker down to the smallest detail. Each program was built in four stages: prepare – present – apply – test. This approach involved professionals (employees) led by professionals (managers), with goals focusing on improving productivity. This methodology was later taken up by Toyotism and more widely by lean management (on this question, see Chapter 3). Originally, the work situation was seen as an opportunity to apply and test the learning "part", which must be assimilated as quickly as possible by the worker. Everything is in place, so that the educational model perfectly embodies a certain vision of work and professional organization. First of all, we find one of the grounds for Taylorism: some people think at the place of others, who will execute such ideas. The engineer is the one who knows what must be learned and how it must be learned. This fundamental distinction between design and production is a reflection of what is at stake in work organization. We learn as we work in the Fordist company. Assigned to a workstation that must be as productive as possible, there is no point in multiplying learning. Here, we learn on-the-job because that is where the source and its purpose are. This situation is sufficiently regulated so that it can be best mastered through learning ad hoc. In fact, it is not so much a question of learning in the situation, but of learning the task and only the task. This approach also makes professionalization impossible – as we understand it today – since it limits training to the task. On the other hand, it is perfectly consistent with the vision of the organization and work of the Fordist company. It is interesting to note that most of the works or testimonies of the time insist on the effectiveness of this pedagogy. The ingredients of a certain form of learning are well mastered: division into sub-tasks and levels, demonstrations followed by testing, correction and immediate re-testing, immediate feedback, etc. This educational model is also that of imitation in its

4 Training within Industry was designed to support the war effort when there was a shortage of manpower. Designed by the American War Department, the method was explained by Charles R. Allen in his book *The instructor, the man and the job*.

poorest sense, the most lacking in perspective; even if we know how to do it, we only know how to do "that". Like a society where social classes tend to make it very difficult to escape from your condition. *Lean management*, which claims its affiliation with the TWI, can sometimes carry the same vision of professionalization reduced to a minimum, as we see in certain American logistics platforms.

7.1.3. *Learning means participating in a professional community in a "peripheral and legitimate" way*

It may seem inappropriate or surprising, after *compagnonnage* and the TWI, to talk about the early work of Lave and Wenger (1991). However, with the idea of legitimate peripheral participation, these authors impose two strong ideas: learning is always social and situated, and it is governed by rules of progressive participation in the métier. Also, even if these works are primarily concerned with describing the processes for acquiring know-how, they largely demonstrate that these are governed by social organizations, rather than being individual cognitive approaches. The professional community is not officially the holder of power, as in the previous examples: hierarchies are expressed more in the form of expertise than managerial power. Its legitimacy is based primarily on experience and mastery of professional gestures, which explains its right to authorize access to knowledge for novices. Organizing work in such a way as to build the career path toward professionalization is a social fact, and this goes without discussion. The conceptual framework of Lave and Wenger (1991) is based on three ideas presented at the beginning of their work[5]:

> "Learning is an integral and inseparable aspect of social practice" (p. 31).
>
> "There is no activity that is not situated" (p. 33).
>
> "Learning is an integral part of generative social practice in the lived-in world" (p. 35).

The authors then draw on five learning stories taken from research on different continents: the midwives of Yucatan (Mexico) (Jordan 1989), the tailors of Liberia (Lave 1982), the quartermasters of the US Navy (Hutchins 1996), apprentice supermarket butchers in the United States (Marshall 1972) and Alcoholics

5 All quotes in this paragraph have been drawn from Lave and Wenger (1991).

Anonymous (Cain 1991). The first three studies and in a certain way that of AA describe the same type of learning and the idea of peripheral participation: progression in the tasks entrusted, important role of the informal and family and community relations (the midwives are almost always the daughters of midwives in Yucatan; West African artisans become apprentices to other families). On the other hand, the study of the butchers' apprentices is different, in that it describes a process which does not run smoothly: "butchers' apprenticeship in contemporary supermarkets often doesn't work" (p. 65), which seems to prevent learning. Marshall insists on the negative role of professional schools, disconnected from the métier and which issue certificates validated by the union, but unrelated to real practices in supermarkets. He then describes on-the-job training periods, where apprentices are confined to repetitive tasks in spaces which do not allow them to observe the work of professionals. The only hope is the arrival of an even more novice apprentice who will justify the move to other, more "legitimate" activities. In this case, the spatial arrangement of labor reflects this character of legitimate peripheral participation: being able to see, observe and be seen are situated social components of learning. These are precisely absent in the case of apprentice supermarket butchers.

This work also allows the authors to structure the concept of community of practice, defined as "a set of relations among persons, activity, and world, over time and in relation with other tangential and overlapping communities of practice" (Lave and Wenger 1991, p. 98). The authors note the variety of configurations: "In practice, the roles of masters are surprisingly variable across time and place" (p. 91). Likewise, the *curricula* (learning or training curriculum), the modes of participation, the balance between the formal and the informal, the role of other apprentices, all of this changes from one social context to another: "a learning curriculum is essentially situated… a learning curriculum is thus characteristic of community" (p. 97). However, it should be noted that "a community of practice is an intrinsic condition for the existence of knowledge not least because it provides the interpretative support necessary for making sense of its heritage... The social structure of this practice, its power relations, and its conditions for legitimacy define possibilities of learning" (p. 98). These concepts would be widely taken up, criticized and enriched in the following years. We will come back to them later because they pave the way to the whole research on workplace learning (WPL).

Concluding this first part, let us note that in different forms, on-the-job professionalization can be analyzed as a set of social processes located and intimately related to relationships of power and domination, which are exercised at a given moment toward the learners. Not only do we learn a profession, but also the social organization of the profession, and the place we must/can occupy in it.

7.2. The individual at the center

To understand the emergence of another vision of on-the-job professionalization, we must take a detour through its rejection. By focusing on the individual, this second model considers that professionalization must take place outside of work situations in order to better master them during the transfer.

7.2.1. *The turning point of individualism and the coronation of the employee actor*

The sociocultural context at the end of the 20th century was marked by what has been called, sometimes a little quickly, the "individualization of society", including the decline of institutions (Dubet 2002), self-invention (Kaufmann 2004), the return of the actor (Touraine 1984) or the search for positive individualism (Singly 1998). The French law of 1971[6] is historically in line with popular education (Besse et al. 2016) and institutionalizes the *individual* right to training. At the same time, the rise of persistent unemployment becomes a strong social marker. At a more or less rapid pace, three complementary ideas set in around vocational training: to fight against unemployment, people must be trained; for the training effort to reach everyone, we must legislate; to finance this effort, we must rely on businesses, the state and, to a lesser extent in France, individuals themselves. In Quebec, the development of texts on adult education was later and clearly marked by an unemployment context, but the orientations were similar: forcing companies to pay to train their employees[7] (Doray and Bélanger 2005). Among the general principles of Quebec's training policy, beyond the state, responsibility concerns "people who benefit from them, their teachers, professional associations, trade unions and socioeconomic organizations, employers and citizens in general" (Québec Ministry of Education 2002, p. 3). In Western countries, the individualization of societies and the fight against unemployment would coexist. The vision of the training would be impacted. This involved training individuals so that they could exercise their right to professional development and to protect them from the risk of unemployment. The "workforce shift" to use Paquet's expression (1993) asserts that "the ultimate goal is the transfer of learning into the work situation" (Gouvernement du Québec, Service de la formation aux adultes, 1989, p. 21; author's translation). However, before this "ultimate goal", training would be removed from work situations and concern individuals learning for themselves, without certainty as to their professional future. Training became a personal commitment taking place outside the constraints of

6 Followed by many other laws and reforms: 2004, 2009, 2014, 2016 and 2018.

7 The law on the development of workforce training (also known as *loi 90* or 1%) requires companies with a payroll of over $250,000 to invest 1% of this mass in training activities.

work. It is necessary to escape the workplace – and its pressures – to devote ourselves to training a few days a year. Companies are required to invest (Meignant 1993) in "human capital" (Becker 1964), to be able to claim a strategic approach to HRM (Guérin and Wils 2002; Cappelletti 2010). This requires taking individuals out of the company so that they can take care of their "skill capital" (Ledru and Michel 1991). The consequences would be visible in terms of educational engineering. As experience was no longer there to learn, we had to design schemes capable of identifying and developing professional skills. A set of roles was thus formed which would mark 40 years of professional training: the individual as the actor of training, the company investor in this human capital, and the "training actors": educational engineers, professional trainers, organizations training, professional centers for adults, financial intermediaries, career observatories, etc. Each country would develop its "little world of training", which ultimately replaced the internal professional order of companies. The economic weight of the training sector[8] is far from negligible in a country like France, which can explain some resistance to change. The training market is based on the purchase of sessions, counted in hours and taking place essentially during working hours and outside the work situation. The public authorities are there to guarantee access to training and support for the most vulnerable. The title of the 2018 law in France "For the *freedom* to *choose* one own professional future" emphasizes this personal freedom. Confronted with the failure of training policies in terms of equal opportunities and the fight against mass unemployment, something regularly observed (Verdier 1995; Santelman 1999, 2006), the response came down to strengthening the individual dimension.[9]

7.2.2. *Learning: developing skills and transferring them*

Competence would be the central notion of this individualization of training. Presented in France as an educational (Le Boterf 1994; Perrenoud 1994), managerial (Aubret et al. 199), social (Zarifian 1999) or political revolution (Zarifian 2009), the skill(s) were a priori contextualized and individualized (Bellier 2004). Few criticisms were formulated (Stroobants 1993) and the concept became widespread (Bouteiller and Gilbert 2005), including in North America and Europe (Winterton

8 The French media frequently use the figure of 32 billion euros, without specifying that it brings together very different elements: educational costs, salaries paid during training, "compulsory" financing from companies for their employees or paid to former OPCAs, state and regional funding for the unemployed, etc.

9 This clearly liberal vision would be modulated by support offers. Unfortunately, the Professional Development Advisor created by the 2014 law remains poorly funded and little used: http://www.cnefop.gouv.fr/rapports-139/cep-4-ans-de-mise-en-oeuvre.html.

2009), where it became the basis of accreditation systems[10]. This shows its importance. However, at the same time, by becoming mostly a management tool, the notion lost its interest: initially, the skills-based approach had to guarantee that real activity was being taken into account. Gradually, it was embodied in the production of general benchmarks which could not grant contextualization. In France, there was also the legal obligation[11] to negotiate on the forward-looking management of jobs and skills (GPEC), which strengthened the institutionalization of the approach. Another confusion, certain skills became transversal (Éducation Permanente 2019), further distancing the skills from real work situations. At the European level, as in the companies, soft skills opened the field to questionable behavioral benchmarks (Bellier 2004). Finally, let us point out that not all skills are equal! As elsewhere, it is the best-off who continue to train in "good skills".

How would professionalization schemes be designed in that context? Work situation and training seemed to conflict. To "acquire skills", we had to return to the adult school, internal or external to companies: the training center, the professional academy and the skills university. All these titles overlapped with the same spatial approach: learning took place outside the work situation. The key lay in the design of meaningful off-the-job learning, expected to guarantee the transfer in a professional situation. Vast literature was published on educational engineering (Hameline 1979; Bellier 1999; Rieunier 2000), taxonomies (Krathwohl et al. 2002; Berthiaume and Daele 2013), educational design (Andrews and Goodson 1980; Cross 1984, 2004; Allen 2006; Spector et al., 2014) and by extension, pedagogic facilitation. Instead of disappearing, the work situation became the place where to practice what was "learned" during training. "Apprenticeship pedagogies"[12] became a sort of quality assurance, but the implementations were varied and, over time, the practical, as well as theoretical question of transfer, came up (Savoirs 2006). Works from English-speaking countries and from Quebec[13] (Butterfield and Nelson 1989; Perkins and Salomon 1989; Tardif 1999) inaugurated that path. Haskell (1997, 2001) proposed a model that was widely adopted and enriched (Roussel 2011). Understanding what makes a learner effectively apply what they have acquired during training to the work situation makes it possible to emphasize the decontextualization/recontextualization processes (Roussel 2011). Ultimately, this vision of professionalization perfectly reflects 30 years of a liberal

---

10 The European Council in Lisbon in 2000 established the ECTS system for all European universities, thus facilitating student mobility.

11 French law on professional training from 2004.

12 See the very complete chapter by P. Maubant on the subject in this work.

13 Let us nevertheless cite the excellent seminal work by Rey, B. (1996). *Les compétences transversales en question*. Editions sociales françaises, Paris.

political model based on individualism (Corcuff et al. 2010): *the individual* as a holder and supplier of skills, *the company* investing in this individual capital and *experts* responsible for maintaining the capital in question. Is this not a sort of mirror of financial capitalism: holders of capital, companies requesting this capital and experts (the banking system) trying to make the system more efficient?

For Carré (2005, 2020), apprenance is first defined as "a lasting set of dispositions favorable to the act of learning in all situations". It is therefore an individual attitude which includes cognitive, affective and conative dimensions. However, noting that this notion is used quite extensively today, he proposes the idea of a change in culture: the emergence of a "learning culture" to replace the educational culture of the 20th century. The theoretical framework is that of a learning social subject, and of a triadic model putting into dynamic relationship individual dispositions or factors, a social context or environment, and behaviors or practices. Carré thus brings together the works of Bandura, Lahire and Léon in an integrative approach. Although the question of professionalization is not the central subject of works on apprenance, this notion invites us to reconsider the schemes in the light of learner agency. "To become professional" would be the exact term, which does not prevent us from insisting on the role of context, peers, practice opportunities, collectives and the facilitation provided by the support professions that Carré summons to replace those of traditional training. The interest and the risk in the notion of apprenance is that it shifts the questions of learning toward individual postures, minimizing the role of schemes designed by others. These are individuals who learn according to their needs, in their concrete situation. Consequently, the role of educators must be profoundly reconsidered.

Box 7.1. *Apprenance, a paradigm shift?*

7.2.3. *The digital revolution and the health crisis: learning and working remotely from the organization*

Professionalization seen from the angle of the possession of individual skills and their transfer on-the-job would be profoundly disrupted by what has been called the "digital revolution" or even the "dazzling transition" (Giorgini 2014). This is for at least three reasons: the use of digital tools has become more widespread, changing access to information but also its virtualization, messaging and other social networks have disrupted collective functioning as well as work and life rhythms (Virilio 2010), and finally, hierarchical relationships are deeply questioned by a more horizontal vision of society (Enlart and Charbonnier 2013). It is an understatement to say that the individual relationship to work has changed, and with it, the construction of professional identities, the commitment to organizations, and the relationship to knowledge and professionalization (Enlart and Charbonnier 2010). The notions of individual performance, efficiency, short time imposed themselves:

under the influence of ICT and e-learning, already present since the 1990s, there was a development of training far from the field – even if blended – under the cover of digital modernity. Professionalization now involves micro-learning, capsules, "grains of knowledge" on the shelf that everyone can consume when and where they deem it useful. Training practices are modified, in complete consistency with a society having immediate access to data. Prospective work on professions increases, emphasizing the disappearance of jobs and the emergence of future professions which are still unknown. This leads to two contradictory reactions: on the one hand, we must professionalize more and more quickly, but, on the other hand, why do it if métiers are constantly changing? Of course, digital technology alone cannot explain the whole of these transformations, but it has represented an impressive phenomenon of acceleration and generalization. Society, and then businesses, started to wonder how to "remain human" (Enlart and Charbonnier 2018). Reflections on professionalization would offer answers to those questions. In this sense, the work of Crawford (2010, 2015) was premonitory: by pleading for a return to the concrete, the manual, the physical world, he rehabilitated a universe of "contacts", opposed to virtual spaces. The more virtual society became, the greater the thirst for informal learning between peers, on the field and outside organizational control. Fablabs, the world of hackers, co-working spaces, third places would emerge as premises where professionalization is done in a different way. The example of AFEST[14] (see Box 7.2) is part of this movement of a return to the field.

Legally legitimate since the French law on professional training of September 2018, AFEST is a pedagogic modality which can be part of a course and invites the development of individual skills in a real work situation. It is based on specific pedagogic support offered to a volunteer learner to sequentially develop the skills targeted with the supervisor. The learner will be observed and then have a reflexivity interview to become aware of the way in which they have worked and what could be done differently. Trainers/facilitators can be part of the company as long as they are trained in the process. After having verified the feasibility and relevance of AFEST, the conditions for success should be noted, including the learner's readiness, the degree of support on the part of the "tutor" who, in addition, does not have a hierarchical bond with the learner. Pragmatic and close to the work context, AFEST involves a change in the pedagogic model and places the company's actors at the center of the schemes. Thus, AFEST clearly illustrates the refocusing on a learner at work, in a particular context, far from catalog training, and the whole managed by the company itself.

Box 7.2. *On-the-job training actions (AFEST)*

14 The French law of 2018 on professional training which recognized "on-the-job training actions" as a training modality came to be known as AFEST (law of September 2018 on professional training).

Digital transformation has provoked a counter-movement: reinvesting physical workplaces, even if this means rethinking spaces (Benedetto-Meyer and Cihuelo 2016; Giorgini 2016). However, a second shock would shake the world of work and thwart this return to reality. The health crisis of 2020–2021 established teleworking as the norm, making relationships with physical spaces increasingly distanced and virtual. Without this crisis, perhaps we would have witnessed a rebalancing between the virtual and the real, other ways of thinking about learning, both remotely but also within groups gathered physically; perhaps a new pedagogy of apprenticeship could have seen the light of day. By forcing many companies to generalize remote working, the Covid crisis also made learning processes more difficult, especially for youth or "newbies" entering a professional sector. Remote tutoring made informal communication difficult, since there is an emotional and affective dimension which develops over the long term after working side by side, in itself a factor of commitment fueling the "desire to learn" (Bourgeois 2018).

We will have to wait a few more years to analyze how these two societal crises will impact on professionalization models, but we can already bet that they will have repercussions on learning models. On-the-job learning is clearly dependent on how these situations evolve, not due to pedagogic or organizational choices, but rather under the impact of a profound societal change.

7.3. Activity at the center[15]

A third current is mainly interested in on-the-job professionalization. The intention is to understand what workers truly do, what their activity really is, even if they do not always know how to express it themselves. The common thread of this research is the analysis of concrete situations, those which allow us to become professional in a specific company. However, today this is regularly changing. The professionalization in question here involves becoming familiar with the reality of the métiers in the field, as accepting their transformation. Focusing on the activity enables us to detect appropriation processes, as adaptation to the context. In a rapidly changing world, which has overcome the great ideologies[16], the profession is learned as citizenship is experienced in a person's family, neighborhood and community, including with new media, such as social networks (Casilli 2010).

15 This section refers to well-documented models in this work. We will be quicker and refer to the relevant chapters.

16 When they exacerbated them in other worlds in extremist forms.

7.3.1. *WPL and communities of practice: an encompassing theoretical framework*

Research on *WPL* has been dominant since the 1990s, particularly in the English-speaking countries. This success is also a source of difficulty, because under the term WPL, there is a broad spectrum in research (Malloch et al. 2011) whose common point is this idea of the "workplace". However, as Illeris (2011) points out, it is a notion that deserves to be debated. Another issue is that WPL studies learning not only in its cognitive dimensions, but also emphasizes its social aspects. For example, when Billett[17] (2001) analyzes the professionalization of apprentice hairdressers, he describes its contextual variables and psycho-cognitive dimensions (Billett 2007), while introducing learning as a social process. So, here is an encompassing epistemology that seeks to establish a connection between all individual, cooperative and organizational registers, to such extent that nothing escapes the workplace. This is where *curricula* must be written, including both formal and informal aspects of learning. The organization resumes its professionalization role in situ, as the workplace. Working, with others, in a professional community is considered key. The work of Lave and Wenger (1991), already cited, played an essential role in the initial structuring of the field. At the beginning, as we have seen, the topic concerned the modes of entry into a profession, through progressive participation in activities, from the most peripheral to the most central ones. Then, it was a question of analyzing the interactions within communities of practice and their roles in professional learning. Research focused on exchanges and negotiations within the community, opportunities to share the meaning of our experience and practices and to build a common repertoire (Wenger 1998; Wenger and Gervais 2005). From this work would also arise a desire to organize communities as a scheme for learning at work. Pragmatic orientations characterize these models which seek, not only to understand, but also to intervene and develop on-the-job professionalization. The collective actor – a limited and if possible self-determined collective – becomes the essential place of each person's training. This development appears at a time when it is becoming increasingly difficult to predict "good training" and employment prospects. It is difficult to claim to design pedagogic schemes that will respond to professional transformations. Relying on communities is, on the contrary, giving ourselves a chance to think about professionalization as closely as possible to these changes, without adopting an overarching vision but delegating, in some way, the professional adaptation processes. The interest in WPL and in communities of practice emerges at a moment which requires managing changes, adapting and inventing new actions without losing the basics, and this in a horizontal rather than a vertical manner. This need for

17 We will not elaborate further on Billett's work, but refer the reader to his chapter.

adaptation is typical of a speculative economy (Gomez 2019) where future performance must break with the past and is immersed in a race for generalized performance. Confronted with this type of injunction, widely relayed by the digitalization of society and the promotion of a few start-ups and other unicorns, professionalization processes seek to be bearers of innovation and efficiency, but also to protect individuals against absurd demands. By recalling that it is *in* and *with* work that skills are built collectively, communities and the WPL play the role of a firewall of contention. These counter-discourses are not the work of experts speaking for others: they are carried by the actors themselves, structured in communities and networks.

7.3.2. *Professional didactics: professional activity, at the heart of schemes*

With WPL and community studies, research refocuses on activity. Far from the training-catalogues intended for the employee "actor of his/her professional career", the focus is on tracking down real work, which requires a detailed understanding of the human *activity* (Dujarier et al. 2016). Ergonomics and professional didactics (Pastré et al. 2006)[18] intend to grasp what this activity covers at/from work. Piagetian-inspired, professional didactics (PD) was initially characterized by a cognitive and individual approach to professionalization. Seeking to identify the "theorems-in-act" and the "invariants" engaged by the operator, PD proposed a research program and a training approach based on putting into words "what goes on in people's minds" while they work. In this sense, it is close to work psychopathology (Clot 1998, 2015, 2017; Dejours 1998, 2010), which attempted to understand what prevents work and causes suffering. Activity is considered as a set of practices that belong only to those who carry them out. Thinking about training is therefore not thinking about it from our knowledge as a trainer or an expert, but from the mental representations and social constructions that emerge closer to the field. Indeed, what all this work would show is that the métier is always learned in-context and through social interactions. To become professional means enabling everyone to participate, but by understanding what they are doing, taking charge of it, owning the occupation, which presupposes that the words of those who practice it – individual or collective words – can be heard and recognized as legitimate. These proposals also have political dimensions, since they question the balance of power in organizations. However, these are not demands on working conditions in the classic sense. It is more about thinking about the quality of work as a means for professionalizing, while respecting the physical and mental integrity of everyone. This research would lead to forms of work organization where everyone is given a voice and where collectives can be formed, including with management

18 We refer to the chapters on the subject in this book.

(Detchessahar 2019), in spaces for discussion. Unlike what was experienced in the 1960s, these exchanges are less political, but tend to focus on work and its quality. They rehabilitate operators, considered the true holders of the métier's knowledge. Ultimately, we are all experts in what we do, even if we do not always know how to put it into words: not everyone has spontaneous access to the "conceptual structure of the situation" and to their own "operational models"! According to this approach, professionalization takes place as close as possible to the field, but with complex theorizing in the hands of researchers. Is this not the reflection of a society which exacerbates the cognitive dimensions, while wanting to be more "horizontal"? Let us also remark that these approaches are the fruit of several French traumas around *burnout*, psychosocial risks and suicides at work. These streams of research clearly asserted themselves as attempts to clean up work organizations and to fight against situations of neglect, which are extensively criticized in society.

7.3.3. *Empowering environments and feedback from the learning organization*

A consequence of this interest in the work situation is the rediscovery of the organization – a virtuous organization, which offers affordances[19] and capabilities. Drawn from the work of Amartya Sen, the concept of *capabilities* offers a vision of on-the-job professionalization built around the interactions between individuals and the organization (Fernagu-Oudet 2018). The proposed distinction between skills and capabilities is similar to the one between acting – through the mastery of know-how – and the power to act. Capabilities develop on both the individual and organizational sides, in the nature of the interactions made possible and captured by workers. Thus, an empowering organization is an enterprise "which offers meaning, leeway, opportunities for development and learning, which makes it possible to increase the autonomy and power to act of all actors" (Falzon 2005). In the same vein, we see a strong revival of interest in the field of learning organization, even if behind this term, there are varied proposals. The importance assumed by the question of work has also led us to question the contexts. We can schematically identify three types of interpretation under the flag of the learning organization. A first current, faithful to Argyris and Schön as well as to Senge, gives a large place to the practices of managers and their ability to question their own activity. These proposals aim to change the culture (taking into account complexity, systemic thinking, and the mid-long term, etc.) "bottom down". A second approach – as that of empowering organizations – focuses on the interactions between individual and organization, and the way the latter can offer resources, affordances, which also amounts to questioning the real leeway and organization of work, particularly

19 Term dear to Billett, initially used by Gibson, a psychologist.

around collaboration opportunities. Professionalization is clearly inscribed in the organization and in its concrete translation, which is the work situation. Here, individual cognitive dimensions are less important than the ability to seize opportunities offered by the organization to transform them into individual resources. These conversion factors ultimately explain a large part of learning. Finally, a third approach focuses on the work situation, and the way in which it can become a learning experience. The view is "micro", pedagogic, situated: it concerns the organization of work, individual and collective. It is a matter of prosaically ensuring that everyone can continue to learn while working. This involves not only the arrangements at the heart of the work situation, but also in the organizational environment (management, HRM, IT management, job prospects, etc.), which makes learning possible (Enlart 2018).

Beyond their differences, these approaches all propose taking into account the individual in their work context in a processual approach. They all consider that it is the interactions between the two that make it possible to think about professionalization. However, these interactions cannot be planned, controlled and structured beforehand. We must therefore rely on greater leeway made possible thanks to rules of the game and flexible work processes. The challenge is to make professionalization attainable, rather than thinking about it in advance. Awareness of the failures of the neo-liberal system, through repeated financial crises, ecological perspectives and budgetary decisions linked to Covid, has changed the situation and made any programmatic approach definitively fragile.

7.4. Conclusion

Our intention was to show that professionalization measures do not only fall within the field of pedagogy, but that they also include visions of humanity at work immersed in a sociopolitical context. In particular, we sought to analyze the relationship that these schemes have with the work situation and the meaning that this relationship has in the context of social relations. This also refers to the role played by organizations and work collectives. While this angle of attack has always been more or less present in research on professionalization, it has often been implicit or rapid, without seeking to make a systematic connection. We have proposed a reading grid to better connect approaches to professionalization and its place in a social order. This analysis allowed us to identify three models, one centered on integration into a social order, the second centered on the autonomous and responsible individual and the third centered on the worker's activity in a particular context. The connections with socioeconomic and cultural contexts sometimes show mirrored models and sometimes counter-proposals. Although this work deserves further investigation, it has made it possible to explore the question of the meaning that professionalization choices can have, and its relationship to the work situation vis-à-vis the societal context where these choices evolve. Concretely, in the period of crisis we find ourselves in, this type of reflection can be useful for situating the professionalization proposals which are sure to arise. Indeed, it will be particularly interesting to continue this work in the years to come, evaluating how the models presented here persist or transform in the light of new forms of organization and relationships at work.

7.5. References

Abdelnour, S. and Méda, D. (2019). *Les nouveaux travailleurs des applis*. PUF, Paris.

Allen, W.-C. (2006). Overview and evolution of the ADDIE training system. *Advances in Developing Human Resources*, 8(4), 430–441.

Andrews, D.-H. and Goodson, L.-A. (1980). A comparative analysis of models of instructional design. *Journal of Instructional Development*, 3(4), 2–16.

Argyris, C. and Schön, D.A. (1978). *Organizational Learning: A Theory of Action Perspective*. Addison-Wesley, Reading, MA.

Argyris, C. and Schön, D.A. (1996). *Organizational Learning II: Theory, Method and Practice*. Addison-Wesley, Reading, MA.

Ashton, D. (2004). Political economy of workplace learning. In *Workplace Learning in Context,* Rainbird, H., Alison Fuller, A., Munro, A. (eds). Routledge, London.

Aubret, J., Gilbert, P., Pigeyre, F. (1993). *Savoir et pouvoir. Les compétences en questions*. PUF, Paris.

Barraud de Lagerie, P. and Sigalo Santos, L. (2019). Les plateformes de microtravail : le tâcheronnat à l'ère numérique ? In *Les nouveaux travailleurs des applis*, Abdelnour, S. and Méda, D. (eds). PUF, Paris.

Becker, G. (1964). *Human Capital: A Theorical and Empirical Analysis, with Special Reference to Education*. NBER-Columbia University Press, New York.

Bellier, S. (1999). *Ingénierie en formation d'adultes*. Editions Liaisons, Paris.

Bellier, S. (2004). La compétence est-elle un concept nouveau ? In *Traité des sciences et des techniques de la formation*, Carré, P. and Caspar, P. (eds). Dunod, Paris.

Benedetto-Meyer, M. and Cihuelo, J. (2016). L'espace dans l'analyse du travail. *La nouvelle revue du travail*, 9. doi: 10.4000/nrt.2859.

Berthiaume, D. and Daele, A. (2013). Comment choisir des méthodes d'enseignement adaptées. In *La pédagogie de l'enseignement supérieur : repères théoriques et applications pratiques*, Berthiaume, D. and Rege-Colet, N. (eds). Peter Lang, Bern, 119–134.

Besse, L., Chateigner, F., Ihaddadene, F. (2016). L'éducation populaire. *Savoirs*, 3(3), 11–49.

Billett, S. (2001). *Learning in Workplace Learning: Strategies for Effective Practice.* Routledge, London/New York.

Billett, S. (2007). *Work, Subjectivity and Learning: Understanding Learning Through Working Life*. Springer, New York.

Billett, S. (2010). *Learning Through Practice*. Springer, New York.

Borjon-Nieto, J.-J. and Lascurain-Fernandez, M. (2016). *Grandes educadores de Mexico y America Latina.* Secretaria de Educacion de Veracruz.

Bourgeois, E. (2018). *Le désir d'apprendre*. PUF, Paris.

Bouteiller, D. and Gilbert, P. (2005). Réflexions croisées sur la gestion des compétences en France et en Amérique du Nord. *Relations industrielles*, 60(1), 3–28.

Butterfield, E.-C. and Nelson, G.-D. (1989). Theory and practice of teaching for transfer. *Educational Research and Development*, 37(4), 5–38.

Cain, C. (1991). Becoming a non-drinking alcoholic: A case-study in identity acquisition. *Ethos: Journal of the Society for Psychological Anthropology*, University of North Carolina, Chapel Hill, 19, 2.

Cappelletti, L. (2010). Vers un modèle socio-économique de mesure du capital humain ? *Revue française de gestion*, 8(8), 139–152.

Casilli, A-A. (2010). *Les liaisons numériques, vers une nouvelle sociabilité ?* Seuil, Paris.

Clot, Y. (1998). *Le travail sans l'homme*. La Découverte, Paris.

Clot, Y. (2015). *Le travail à cœur*. La Découverte, Paris.

Clot, Y. (2017). *Travail et pouvoir d'agir*. PUF, Paris.

Condorcet (1792). *Rapport et projet de décret relatifs à l'organisation générale de l'instruction publique, Présentation à l'Assemblée législative : 20 et 21 avril 1792.* Bibliothèque de l'Assemblée nationale, Paris.

Corcuff, P., Le Bart, C., de Singly, F. (2010). *L'individu aujourd'hui. Débats sociologiques et contrepoints philosophiques.* PUR, Rennes.

Crawford, M.-B. (2010). *Eloge du carburateur. Essai sur le sens et la valeur du travail.* La Découverte, Paris.

Crawford, M.-B. (2015). *Contact, pourquoi nous avons perdu le monde et comment le retrouver.* La Découverte, Paris.

Cross, N. (1984). *Developments in Design Methodology*. Wiley, Chichester.

Cross, N. (2004). Expertise in design: An overview. *Design Studies*, 25(5), 427–441.

Dejours, C. (1998). *Souffrance en France – La banalisation de l'injustice sociale.* Seuil, Paris.

Dejours, C. (2010). *Le facteur humain*. PUF, Paris.

Detchessahar, M. (2019). *L'entreprise délibérée : refonder le management par le dialogue*. Edition Nouvelle Cité, France.

Dewey, J. (1916) [1997]. *Democracy and Education: An Introduction to the Philosophy of Education*. The Free Press, New York.

Dominicé, P. (1992). *L'histoire de vie comme processus de formation*. L'Harmattan, Paris.

Doray, P. and Bélanger, P. (2005). Vers une nouvelle régulation de la formation continue dans la société Québécoise. In *La formation professionnelle continue*, Guyot, J.-L. and Mainguet, C. (eds). De Boeck Supérieur, Louvain La Neuve.

Dubet, F. (2002). *Le déclin des institutions*. Le Seuil, Paris.

Dujarier, M.-A., Gaudart, C., Gillet, A., Lénel, P. (2016). *L'activité en théories. Regards croisés sur le travail*. Octares Editions, Toulouse.

Education Permanente (2019). *Partir des compétences transversales pour lire autrement le travail*. Hors-série AFPA.

Enlart, S. (2018). *De la formation à la Learning Company*. Eyrolles, Paris.

Enlart, S. and Charbonnier, O. (2010). *Faut-il encore apprendre ?* Dunod, Paris.

Enlart, S. and Charbonnier, O. (2013). *A quoi ressemblera le travail demain ?* Dunod, Paris.

Enlart, S. and Charbonnier, O. (2018). *Société digitale : comment rester humain ?* Dunod, Paris.

Falzon, P. (2005). Developing ergonomics, developing people. *8th South-east Asian Ergonomics Society Conference SEAES-IPS*, Denpasar, Bali.

Fernagu-Oudet, S. (2018). *Organisation et apprentissage : des compétences aux capabilités*. CREF, Paris.

Freire, P. (1968–1974). *Pédagogie des opprimés*. Editions Maspero, Paris.

Giorgini, P. (2014). *La transition fulgurante*. Bayard, Paris.

Giorgini, P. (2016). *Au crépuscule des lieux*. Bayard, Paris.

Gomez, P.-Y. (2019). *L'esprit malin du capitalisme*. Desclée de Brouwer, Paris.

Guérin, G. and Wils, T. (2002). La gestion stratégique des ressources humaines. *Gestion*, 2(2).

Hameline, D. (1979). *Les objectifs pédagogiques*. ESF, Paris.

Haskell, R.-E. (1997). *Reengineering Corporate Training*. Quorum Books, London.

Haskell, R.-E. (2001). *Transfer to Learning*. Academic Press, London.

Heinich, N. (1993). *Du peintre à l'artiste : artisans et académiciens à l'age classique*. Editions de Minuit, Paris.

Hutchins, E. (1996). Learning to navigate. In *Understanding Practice*, Chaiklin, S. and Lave, J. (eds). Cambridge University Press, New York.

Illeris, K. (2011). *Kompetence.* Samfundslitteratur, Frederiksberg.

Jordan, B. (1989). Cosmopolitical obstretics: Some insights from the training of traditional midwives. *Social Science and Medicine*, 28(9), 995–1044.

Kaufmann, J.-C. (2004). *L'invention de soi.* Armand Colin, Paris.

Krathwohl, D.R., Bloom, B.S., Masia, B.B. (1964). *The Taxonomy of Educational Objectives.* David McKay, New York.

Lallement, M. (2015). *L'âge du faire. Hacking, travail, anarchie.* Le Seuil, Paris.

Lave, J. (1982). A comparative approach to educational forms and learning processes. *Anthropology and Education Quarterly*, 13(2), 181–187.

Lave, J. and Wenger, E. (1991). *Situated Learning, Legitimate Peripheral Participation.* Cambridge University Press, Cambridge.

Le Boterf, G. (1994). *De la compétence, essai sur un attracteur étrange.* Les Éditions d'Organisation, Paris.

Ledru, M. and Michel, S. (1991). *Capital-compétence dans l'entreprise : une approche cognitive.* ESF Éditions, Paris

Malloch, M., Cairns, L., Evans, K., O'Connor, B. (2011). *The Sage Handbook of Workplace Learning.* Sage Publications Ltd, London.

Marshall, H. (1972). Structural constraints on learning. In *Learning to Work,* Geer, B (ed.). Sage Publication, Beverly Hills, CA.

Meignant, A. (1993). *Manager la formation.* Editions Liaisons, Paris.

Paquet, P. (1993). L'évolution des politiques canadiennes et québécoises de formation des adultes depuis 1960. In *Enjeux actuels de la formation professionnelle*, Dandurand, P. (ed.). Institut québécois de recherche sur la culture, Quebec.

Pastré, P., Mayen, P., Vergnaud, G. (2006). La didactique professionnelle. *Revue française de pédagogie*, 154, 145–198.

Perkins, D.-N. and Salomon, G. (1989). Rocky roads to transfer: Rethinking mechanisms of neglected phenomenon. *Educational Psychologist*, 24, 113–142.

Perrenoud, P. (1994). Compétences, habitus et savoirs professionnels. *European Journal of Teacher Education*, 17(1–2), 45–48.

Perrot, M. (1996). La jeunesse ouvrière : de l'atelier à l'usine. In *Histoire des jeunes en occident : l'époque contemporaine*, volume 2, Levi, G. and Schmitt, J.-C. (eds). Editions du Seuil, Paris.

Québec Ministry of Education (2002). Government Policy on Adult Education and Continuing Education and Training [Online]. Available at: www.education.gouv.qc.ca/fileadmin/site_web/documents/dpse/politique_a.pdf, [Accessed 19 June 2024].

Rey, B. (1996). *Les compétences transversales en question*. Editions sociales françaises, Paris.

Rieunier, A. (2000). *Préparer un cours*, volumes 1 and 2. ESF, Paris.

Roussel, J.-F. (2011). *Gérer la formation : viser le transfert*. Guérin, Quebec.

Santelmann, P. (1999). *La formation professionnelle continue : bilan et perspectives. Problèmes politiques et sociaux*. La Documentation française, Paris.

Santelmann, P. (2006). *La formation professionnelle continue*. La Documentation française, Paris.

Savoirs (2006). *Transfert de compétences : a quoi sert la formation en entreprise ?* L'Harmattan, Paris.

Senge, P. (1992). *The Fifth Discipline: The Art and Practice of the Learning Organization*. Century Business, London.

Service de la formation aux adultes (1989). *Le transfert des apprentissages dans une démarche de formation sur mesure*. Ministère de l'Enseignement supérieur et de la Science, Gouvernement du Québec, Quebec.

Simon, J. (1867). *L'ouvrier de huit ans*. Librairie internationale – Lacroix, Verboeckhoven et Cie, Paris.

Singly, F. (1998). Individualisme et lien social. *Lien social et politiques*, 39, 33–45.

Spector, J.-M., Merrill, M.-D., Elen, J., Bisho, M.-J. (2014). *Handbook of Research on Educational Communications and Technology*. Springer, New York.

Stroobants, M. (1993). *Savoir-faire et compétences au travail, une sociologie de la fabrication des aptitudes*. Editions de l'Université de Bruxelles, Brussels.

Tardif, J. (1999). *Le transfert des apprentissages*. Editions Logiques, Montreal.

Touraine, A. (1984). *Le retour de l'acteur*. Fayard, Paris.

Verdier, É. (1995). Politiques de formation des jeunes et marche du travail. La France des années quatrevingts. *Formation Emploi*, 50, 19–40.

Virilio, P. (2010). *Le grand accélérateur*. Galilée, Paris.

Wenger, E. (1998). *Communities of Practice: Learning, Meaning and Identity*. Cambridge University Press, Cambridge.

Wenger, E. and Gervais, F. (2005). *La théorie des communautés de pratiques : apprentissage, sens et identité*. Presses de l'Université Laval, Quebec.

Winterton, J. (2009). Competence across Europe: Highest common factor or lowest common denominator. *Journal of European Industrial Training*, 33(8/9), 681–700.

Zarifian, P. (1999). *Objectif compétence : pour une nouvelle logique*. Editions Liaisons, Paris.

Zarifian, P. (2009). *Le travail et la compétence : en entre puissance et contrôle*. PUF, Paris.

8

Learning in the Workplace: Recurrent and Emerging Conceptions and Practices

Stephen BILLETT
Griffith University, Brisbane, Australia

Learning through work stands as the most long-standing, widespread and central means by which occupational knowledge has been developed and advanced across human history and culture (Singleton 1989, pp. 3–30; Pelissier 1991; Roodhouse 2007; Billett 2014). Even in the current era when much initial preparation and ongoing occupational development is being undertaken through programs organized through or in educational institutions, workplace experiences remain essential elements of development (Cooper et al. 2010; Dean and Sykes 2021). However, in "schooled societies" there persists a view that workplaces are inferior and ill-suited environments for learning (Moore and Klein 2020). Indeed, they have long been viewed primarily as an adjunct to experiences in educational institutions or mainly as opportunities to practice, apply and refine what has been learnt in those institutions (Marsick 1988; Marsick and Watkins 1990). The realization now, across a range of occupational fields and tiers of education, that in their own right these physical and social settings provide experiences that are essential to that learning and development is causing a reappraisal of their contributions to initial occupational preparation and ongoing development across working life (Frost et al. 2010; Dornan 2012). This interest is prompting more encompassing and impartial considerations about the educational purposes to which workplace learning experiences can be directed and how those experiences can be optimized in achieving those purposes. This extends to including how they can be augmented,

Encyclopedia of Professionalization,
coordinated by Didier DEMAZIÈRE and Richard WITTORSKI.

engaged with effectively and integrated with experiences provided in educational settings (Henderson and Alexander 2011, pp. 131–148; Kothari and Wathen 2017). These considerations also extend to positioning workers as active and ongoing learners and innovators in and through their workplaces (Halvarsson Lundkvist and Gustavsson 2018; Anvik et al. 2020; Billett et al. 2021, pp. 219–244).

This chapter seeks to overview the evolving developmental purposes for which workplace experiences are being engaged and to discuss contributions of learning through everyday work activities and how these experiences are augmented by the structuring (i.e. workplace curriculum) and activities intentionally aimed to promote effective learning (i.e. practice pedagogies). This discussion draws on accounts of long-standing and emerging pedagogic practices, including the use of simulations and diverse ways in which experienced practitioners can support learning in an era of digitalization (Harteis 2018b), including through digital mediums (Billett 2021). Hence, it contributes to the current broadening consideration of what constitutes curriculum practices associated with workplace experiences and the means now occurring by which those experiences can be enriched and augmented. Moreover, as individuals' learning is ultimately premised on their mediation of those experiences (i.e. their construal and construction of what they know, can do and value), there is also a growing focus on how their personal epistemologies (i.e. what they know, can do and value) shapes that learning (Billett 2009). This includes how these personal capacities are developed through and come to engage with these experiences, how they exercise their intentionality and an elaboration of their learning processes in circumstances outside of structured "school-like" experiences. It is these evolving issues that are engaged with and discussed here, commencing with a consideration of purposes for learning through work. Then, the means by which those experiences can be augmented and enriched are deliberated upon, as are the premises for understanding and promoting individuals' engagement and learning through what is afforded them in and through work.

8.1. Purposes of learning through work

The key purposes for learning through work are threefold: developing the capacities to be employed in a specific occupation, sustaining those capacities across working life and contributing to workplace innovation and change. The first and perhaps most long-standing purposes associated with learning through work are the initial development of occupational capacities and their further development to meet specific workplace requirements in response to changes in the requirements for occupational competence and workplace needs (Raizen 1991; Berryman 1993). These purposes are particularly emphasized in considerations of young people who can learn their preferred occupations through engagement in work processes

– throughout human history, occupations and cultures (Singleton 1989, pp. 3–30; Rogoff 1990; Marchand 2008; Gowlland 2012; Lancy 2012). This engagement of young people in learning through work activities has become institutionalized in efforts to enhance the quality of learning experiences and generate greater consistency and outcomes through workplace experiences and in efforts to secure sufficient numbers and kinds of skilled workers (Ainley and Rainbird 1999; Deissinger and Hellwig 2005; International Labour Organisation 2015; Esmond 2020). So, the perennial concerns associated with preparing young people for employment in productive work activities have long been, and remain, central to a focus on learning experiences in workplaces. These arrangements are manifested in different ways across jurisdictions through historical events and cultural emphases (Greinert 2002; Gonon 2009; Deissinger and Ott 2016, pp. 268–296; see also Chapter 9).

However, in recent times, concerns have grown concerning the need for working age adults to remain employable and current in their occupational capacities and in their utility for their particular workplaces across lengthening working lives (Organisation for Economic Cooperation and Development 2006). This has arisen because of the constant changes in and transformation of occupational requirements and changing work practices and means of engaging in work. Meeting specific workplace needs has led to a growing focus on the ongoing development of workers' occupational capacities across lengthening working lives (Organisation for Economic Cooperation and Development 1996), including the formulation of assessments of adult competence in workplace problem-solving and technology (Organisation for Economic Cooperation and Development 2013). This imperative of continuous learning for work–life currency has led to identifying ways of securing scalable processes for entire workforces and particular requirements for specific categories of workers. Added here are concerns about the efficacy of professional development or continuing education and training programs that are remote from actual practice, leading to growing considerations of how this development can occur through practice (Frost et al. 2010; Noble et al. 2019). Consequently, there is growing attention to how this development can occur in and through the circumstances of practice (Billett and Newton 2010, pp. 52–65; Eppich et al. 2016, pp. 47–73). As with initial occupational preparation, a key imperative is for this development to occur as part of everyday work activities and interactions, as it is this authenticity of experiences that provides optimal learning experiences. Hence, considerations of how this ongoing learning can occur through existing work processes is becoming of greater interest. Yet, in some instances it is necessary to provide substitute or simulated work activities when there are potential risks to others or to work processes and equipment. Here, simulated activities are often linked to direct work performance, such as in commercial pilot upgrades and annual assessments through briefings and debriefings. Equally, there are concerns about how such experiences can be linked to actual performance within workplace tasks. Hence, there is a growing focus on workplaces as sites of ongoing occupational development.

It is an emerging purpose for these learning experiences to contribute to workplace innovations and changes in work practice. A key realization is that workplace change is premised upon workers' learning to initiate, trial, implement and sustain workplace innovations (Ellström 2001; Hoyrup et al. 2012) and also that there is an interdependence between those innovations and workers' learning (Billett et al. 2018; Anvik et al. 2020). This concern brings together two broad and important societal imperatives: (i) individuals sustaining their employability across working lives as work performance requirements change; and (ii) the constant innovation and change required to meet the evolving societal, community and individual needs that occupations provide. While there is nothing particularly new about this requirement of responding to change through innovations and occupational and workplace practices being initiated and enacted by practitioners (Epstein 2005), it is being accentuated in the current era. This is because many of these changes appear to be greater in amplitude and more frequent in the current era. Hence, the requirement for constant learning and development to meet those needs. Indeed, the Program of International Assessment of Adult Competence indicates that workers of all kinds and classifications are engaging in nonroutine problem-solving on a regular basis in and through their work (Billett 2015).

So, the long-standing and emerging purposes for learning in and through workplace experiences are associated with initial development of occupational capacities, further extending and advancing those capacities in response to changing occupational workplace requirements and the growing requirement for workplace innovations.

8.2. Utilizing, enriching and augmenting workplace learning experiences

How work experiences can be used to initially develop and extend occupational capacities and workplace innovations requires considerations of how that learning can arise and best be supported, augmented and guided. A key principle for considering their use for these purposes is that, whenever possible, they should occur as part of everyday work activities and interactions, as activity structures cognition (Rogoff and Lave 1984). This principle is founded on both pragmatic and pedagogic bases. The pragmatic imperative is that both public and private enterprises' activities and interactions are largely directed toward achieving the goods and services that they generate. Broad engagement by entire workforces in any means other than at or through work is unlikely to be possible, funded or exercised widely. There will, of course, be some exceptions, such as when new legislative requirements are introduced requiring specific training. Even then, usually, associations with work practice are required for those initiatives to be embedded in practice. Yet, with some exceptions (e.g. military and emergency services), the priority for the existence and continuity of these enterprises is based on their generating goods and services, not on developing workers' skills, which is a lower tier concern. Hence, when considering provision of experiences to assist that learning in and through work, they are usually best found within the circumstances of practice, rather than through the orthodox "training program" options.

Pedagogically, however, it is through engagement in these activities and interactions that much of the learning required for occupational practice can best be generated. Indeed, an important and distinctive attribute of workplace learning experiences is that they provide authenticity in terms of the goals for, activities needing to be undertaken, and the understandings and practices needing to be learnt through engaging in those activities and achieving those goals (Rogoff and Lave 1984; Billett 2001). That is, unlike the hybrid experiences created for the purposes of education or learning in educational institutions (e.g. schools, vocational colleges and universities) and then in needing to transfer this learning to workplaces, workplaces provide access to the kinds of activities that must be learned. This requires individuals, sooner or later, to participate in these experiences. The key questions are how that participation should occur and in what order, and in what ways, it is most likely to support effective learning. Addressing these questions requires a different way of considering support and guidance for learning than is usually entertained in the educational discourse; this discourse primarily concerns the hybrid experiences provided by the educational institution and teachers are the key focal point. It is worth noting that across human history the most common modes of occupational development have been through apprenticeship as a mode of learning (and not as a model of education) (Billett 2016). Apprenticeships as a mode of learning is premised on novices' active construction of occupational knowledge by gaining insights and capacities to practice the occupation through their own efforts of observation, imitation and practice. The locus for that learning resides with them: the apprentice (Rogoff 1990). Indeed, the very meaning of the word apprentice is *apprehendre* – to grasp or take the knowledge (Webb 1999, pp. 100–110). Yet, it is in the circumstances of practice that this active construction of knowledge occurs in ways that have powerful cognitive, procedural and dispositional legacies (Rogoff and Lave 1984). It is thus helpful to consider how learning can arise through everyday work activities and how those experiences can be guided, supported and augmented to achieve the goals of initial or further occupational development.

However, before doing so, it is worth briefly stating that the considerations here are not to diminish, undermine or deny the importance of educational experiences in educational settings, or the contributions made by teachers. These institutions have been established for specific purposes and teachers have been trained to augment, guide and support learning. The key question for the educative role of these institutions in the initial and ongoing development of occupational competence and achieving workplace innovations is found in their hybridity. That is, they been created to achieve these specific purposes and it is important to distinguish for what kinds of learning and in what ways these hybrid settings for learning and the processes underpinning them can be effective in their own terms and can be integrated with experiences found elsewhere, such as in workplaces. This, however, is a separate discussion which is beyond the scope of this brief chapter.

8.3. Learning through occupational practice

As acknowledged by learning theorists over the last hundred years at least, humans learn all the time through the thinking and acting that comprises their experiences (Baldwin 1898; Piaget 1971). Whether drawing upon either social (Vygotsky 1978) or individual (Fodor 1983) constructivism, the common conclusion is that humans are active-meaning makers in responding to what they experience (von Glasersfeld 1987, pp. 3–17). This process is perhaps at its most potent when arising in the pursuit of goal-directed activities that are socially and personally salient (Tobias 1994), that is, when individuals direct what they know (i.e. their conceptual knowledge), can do (i.e. their procedural capacities) and value (i.e. their beliefs and dispositions) in an intentionally goal-directed way to achieve an intended outcome (Malle et al. 2001, pp. 1–26), such as the completion of the work tasks.

The process exercises human intentionality, as it requires that what is known is applied to understand a particular task, problem to be solved or need to be addressed, and then options for achieving that goal are considered and the most appropriate course of action is selected and then monitored through its enactment and judgments made about its success in achieving the intended goal. The exercise of these processes both draws upon and extends what individuals know, can do and value, which, by any other name, is learning. That is, changes arise in individuals' conceptual, procedural and dispositional knowledge through these processes. This explains why so much emphasis is placed in cognitive accounts of learning on problem-solving as a means by which learning can be realized and individuals' knowledge can be appraised (see also Chapter 7 of this volume).

The salient point is that the everyday activities and interactions that workers engage in across their working life are generative of learning, exercise the maxim "activity structures cognition" (Rogoff and Lave 1984); that is, the kinds of activities we engage in, albeit being personally dependent, have a legacy that is referred to as learning. For instance, engaging in routine problem-solving activities stands to reinforce, refine and extend what we know, whereas nonroutine activities are generative of new knowledge, transforming what we know, can do and value. All kinds and classifications of workers engage in both routine and nonroutine problem-solving on a very regular basis (Organisation for Economic Cooperation and Development 2013). It is through these experiences that they learn: with deeper learning arising from the workers' own efforts compared to when they are assisted by more experienced co-workers. This OECD study permits a detailed analysis of large numbers of workers' learning in all classifications of work, age, backgrounds and across a range of countries.

So, the important point here is that in most cases, the learning required for occupational practice and work can arise through participation in everyday work activities and interactions. That is, this learning arises ordinarily through work as engagement in goal-directed activities in their workplace. This long-standing learning process has been described as mimetic learning that comprises the process of observation, imitation and practice (Billett 2014). This process seems to be central to how learning arises across human history and across individuals' lives. Importantly, it is the physical and social environment of the workplace that affords the basis for this learning in terms of models, artifacts and interactions, and it is being pressed into goal-directed activities that allows purposes to be formed and pursued (Lave et al. 1984, pp. 76–94). Hence, the authenticity of these activities affords particular support for and contributions to that learning which is described as inter-psychological between the person and the social world beyond them. The knowledge that is required for occupational competence and workplace performance does not come from within the individual: it must be accessed and secured from outside of it (inter-psychologically, or beyond the skin (Wertsch and Tulviste 1992)). Workplace activities and interactions provide this access and the processes of appropriation of that experienced. This is what makes the authentic experiences in workplaces so essential for learning occupational practices and workplace requirements.

It is also important to emphasize here that beyond what is afforded by workplace activities and interactions is how these are mediated by workers' personal epistemologies (Billett 2009). Individuals engaging in and deriving meaning through observation participate in the process of imitating, even in its imagined form. These are mediated by their personal epistemologies comprising what they know, can do and value, thereby emphasizing the key role of the active participation, engagement and generation of outcomes by individuals: the construing and construction of what is being experienced. So, although in educational institutions the teacher has a key role in mediating the knowledge to be learnt, in work settings the learning processes are largely a product of individuals' mediation of their learning, albeit assisted and guided by more experienced co-workers and others who provide models to be observed and emulated.

Yet, as with any other circumstance, there are strengths and limitations associated with learning through practice. For instance, as noted, much but not all of the knowledge required for occupational competence can be accessed and learnt through engagement in everyday activities and interactions in the practice of those occupations. However, there is much that cannot be secured through discovery efforts alone. Consequently, it is necessary to identify the ways in which that knowledge can be made accessible and engaged with by novices. A key consideration to make workplaces more effective learning environments and to develop stronger learning outcomes is to identify how learning can be supported, enhanced and extended, which is addressed in the sections below.

8.4. Supporting, augmenting and guiding learning at work

As noted, the processes of learning through work are long-standing and well established across human history. Across that time, there are three key premises for guiding, supporting and augmenting that learning. These are (i) the sequencing and organization of workplace experiences, (ii) the practices within the workplace that have particular pedagogic potency and (iii) the active engagement of workers. In this section, the first two of these premises are addressed in terms of the practice curriculum and practice pedagogies. The practice curriculum comprises the sequencing, duration and progression in the activities and interactions through which individuals can incrementally and effectively develop the capacities to practice an occupation (Billett 2006; Rintala and Nokelainen 2019), that is, the organization of experiences that can be engaged with so that the individuals come to access, engage with and appropriate the understandings, procedures and dispositions they need for their work (i.e. the learning curriculum) (Lave 1990, pp. 259–286). Many workplaces have arrangements that comprise the sequencing of engaging in activities that beginners, newcomers or outsiders initially participated in before being given key responsibilities. In the case of known novices, the pathway is often from engaging in activities that demand lower levels of skills and other essential capacities and which have little consequences, should errors be made by them (Lave 1990, pp. 259–286; Goller et al. 2019). Sometimes, these arrangements reflect pragmatic concerns associated with access to equipment, resources or materials. For instance, apprentice potters first observe how experienced potters manipulate the clay as the potter's wheel turns and then try to imitate what they have observed through practice when the wheels are not being used for productive purposes (e.g. during meal breaks, in the evenings) (Singleton 1989, pp. 3–30). They try to accomplish increasingly mature approximations of what they have observed to secure and increasingly enhance procedural skills in shaping the pots. Sometimes, this development is assisted by experienced potters placing their hands upon those of the novices to model and guide their manipulation of the clay as it spins on the wheel. This is an example of pedagogic practices supporting learning through work. It comprises the novice developing mental models of goals and processes, gaining access to the potter's wheel and practicing and securing incremental improvements, sometimes directly guided by an expert, but with much of the initial efforts being returned to the clay bin for reuse in production (Singleton 1989, pp. 3–30). Underpinning all of this is the central role of novices as active learners. Yet, there are a range of practice pedagogies that can support, extend and augment this active learning.

A range of such pedagogic practices are enacted as part of everyday work activities that can enrich and augment that active learning. These include the use of stories or narratives (Jordan 2011; Wallin et al. 2020), experienced workers verbalizing their mental processes while conducting tasks (Gowlland 2012), direct guidance by more expert partners (Noble et al. 2017), having access to partially completed jobs (Makovichy 2010), the use of objects to explain or illuminate something (Pelissier 1991) and use of mnemonics (Rice 2010) to assist remembering and recalling particular concepts. Then, there are also well-established processes associated with procedural development, such as coaching and guided learning (Pylväs et al. 2018; Lui and Choy 2020; see also Chapter 9).

There are also workplace activities and interactions whose qualities have the potential to be inherently effective learning experiences. These are referred to as pedagogically rich activities (Billett and Noble 2020) that have particular qualities. First, they are goal directed and richly contextualized in work activities. Second, they include a discussion about the issues or problems to be addressed and potential options for how they can be addressed. Third, they require participants, who are often experts, to articulate their perspectives on the goals to be achieved, the range of options for processes to be enacted to respond to those goals and deliberations about which pathway should be taken and how the goals will be evaluated. Fourth, there is an environment in which a fulsome exchange of ideas and considerations of them can proceed. While these may seem a daunting set of requirements, these kinds of activities occur in a range of common forms. For instance, production meetings, team meetings, handovers in health and social care, case discussions and evaluation processes, and forms of event or activity briefings and debriefings often ordinarily meet these requirements. Handovers with nursing staff and allied health are good examples here, as are ward rounds in hospital settings (Bakon et al. 2017; Billett et al. 2017, pp. 207–220).

For instance, in early morning allied health handovers, each patient's case is briefly discussed by members of the cross-disciplinary team, including their specific healthcare issue or issues, the treatment being provided, its effectiveness to date and treatment across the coming day and responsibilities for aspects of the patient's care. These discussions are led by a senior clinician and include perspectives, insights and contributions from the range of healthcare practitioners attending, including social workers, dietitians and other specialist practitioners. These discussions represent important opportunities for novices and more experienced health and social co-workers to understand the complexity of these patient cases and to listen to, evaluate and make judgments about the range of options being presented and those being selected for the particular patient's care. These kinds of richly contextualized learning experiences lead to high levels of indexicality (i.e. the structuring and ability to recall knowledge). So, these activities generate what may be referred to as memorable and recallable experiences. Yet, to be effective, these pedagogically rich activities do necessarily draw upon what individuals know, can do and value and prompt them to engage effortfully and to extend their knowledge through actual or imagined participation in that process. So, while these events are pedagogically rich in themselves, through the cases presented and the care options proposed, evaluated and selected, these deliberations are nothing more or less than an invitation for individuals to engage and learn.

A related and emerging basis for supporting and guiding learning through work is also found in activities such as these handovers, that is, the requirement to utilize the pedagogic potential of digitalized artifacts and work formats. In these handovers, electronic workstations are used to provide an array of information about the patients in text, graphical and imagery forms. These are often the key point of reference in discussions, and many of the presentations can represent complex scenarios comprising multiple sets of information that need to be understood. With the increase in the use of digitalized forms of work across a range of occupations (Harteis 2018b), there is a growing need for this knowledge to be learnt in and through work and these technologies often have the capacity to support that learning through symbols, illuminating that which is not possible to be experienced directly (Harteis 2018a, pp. 85–97; Billett 2021). So, these emerging technologies present both kinds of knowledge to be learnt, but also may well offer the means to support the learning.

Importantly, these kinds of activities occur every day in workplaces and offer potentially rich learning experiences that are authentic and grounded and that can lead to robust learning outcomes, premised on two related factors: first, these activities being enacted in an environment that is open to discussion, consideration of options and evaluation of what is being proposed and how it will proceed; second, these activities being engaged with effortfully and in effective ways by learners. If workers view these only as meetings to get through the working day rather than as opportunities for learning, then the learning outcomes are less likely to be rich.

So, a key consideration for promoting learning through work is to identify the kinds of activities and interactions occurring as part of everyday work activity and consider how these can be enacted to afford rich learning experiences, then advising and supporting workers to engage in them in ways that secure robust learning outcomes. Ultimately, however, it is the participation in and exercise of effortful engagement in these activities and interactions that will determine the quality of learning processes and outcomes.

8.5. Promoting and engaging workers' personal epistemologies

As emphasized above, individuals such as workers are meaning makers: they are the ones who learn and develop through the process of experiencing. Yet, this "experiencing" is also active, as humans necessarily need to be discerning in those suggestions from the social world with which they engage. It is necessary to rebuff or ignore many unhelpful social suggestions to focus on that which is important and relevant to them (Valsiner 1998). Central here, as foreshadowed, is the means by which workers take up the invitations afforded to them by the experiences they can encounter in the workplace, the forms of support and assistance available to them and engagement in activities which offer specific potential for learning. However, more than fulsome and full-blooded engagement are also some personal epistemological practices that are likely to assist that learning. The importance of mimetic learning (i.e. observation, imitation and practice) (Billett 2014) is prime among these, representing a foundational approach to how learning of occupational practices seems to have occurred across human history, largely in the absence of educational efforts by others. It is exemplified by the original concept of apprenticeship, which was active engagement and construction by the novice whose role it was to apprehend or seize the knowledge, rather than having it presented to them (Webb 1999, pp. 100–110). So, this active engagement should not be seen as mindless appropriation or passive socialization. Instead, the process is one of engaged, critical and evaluative participation in practice, not the least because this process includes assessing the viability of what is being experienced. Indeed, Valsiner (1998) proposed that these processes can be described as either mastery, in which individuals uncritically accept what is being proposed to them as they are aligned with their earlier experiences or beliefs, or appropriation, in which they critically appraise and construct their knowledge accordingly, most likely extending and/or consolidating it further.

Other personal practices associated with learning through occupational practice are advanced through a range of disciplines including human development accounts that refer to the importance of forming effective working relationships with others, described as ontogenetic ritualization (Tomasello 2004) which permits learning with others by establishing bases for interactions. Also, the concept of deliberate practice – repeated intentional efforts to hone and improve performance – is seen as a key quality of expertise (Ericsson 2006, pp. 685–705; Avby 2016, pp. 75–90). All of these processes stand to mediate not only the initial learning of occupational knowledge, but also its ongoing development through practice and across working life. It is the conscious engagement in real or imagined approaches to problems and challenges that serves to develop further and refine and hone what individuals already know, can do and value. Yet, the unresolved question is how these attributes can best be developed and sustained by workers.

One quality emerging from anthropological studies is the importance of the novice actively wanting to learn and assenting to be engaged in a process of learning (Rose 1990; Mishler 2004; Sennett 2008). For instance, in Japanese communities producing high-quality teapots, it was not sufficient for a child to become an apprentice simply because they grew up in a family of potters. They had to assent to wanting to become a potter (Singleton 1989, pp. 3–30). Similarly, young people in nomadic families in Kyrgyzstan who wanted to become yurt makers, blacksmiths, or traditional storytellers had to assent to leave their family and live with families who practiced those occupations (Bunn 1999, pp. 74–85). Much but not all of this subjectivity can be related to the standing of the occupation within the community and individuals' associations with it. That is, occupations that are seen as being high status and valued may well have a particularly attractive social dimension which may or may not be aligned with the individuals' capacities. Yet, other accounts referred to the importance of the formation of individuals' subjectivity or sense of self, including the processes of identifying with a particular occupation arising through local engagement, satisfaction with work roles and emerging subjectivity as a practitioner (Chan 2013).

In an era of high aspiration and strong preference for clean work that can be labelled professional, there are significant challenges associated with engaging young people in occupations and sustaining that interest across working life that are categorized in other ways and, importantly, posterior to the professions. All of this represents a significant challenge for the development of many occupational capacities as these require effortful engagement by those who learn and then will come to practice them.

8.6. Recurrent and emerging conceptions and practices: learning through work

Proposed in this chapter is the need for workplaces to be seen as social and physical environments that can serve a range of educative purposes associated with individuals' initial and ongoing development across working life and the ability to bring about change in those workplaces. Those purposes can be achieved in part through the normal everyday activities occurring in and through work. Those activities present problems and challenges that are required to be resolved as part of everyday work activities. That resolution is generative of learning, albeit in different ways for workers with different levels or kinds of readiness. Importantly, tasks that are familiar and have been engaged with previously offer the prospect of refining, reinforcing and honing what workers know, can do and value, while novel problems and challenges offer the prospect of developing further and extending workers' knowledge, although sometimes requiring specific forms of support to assist them to achieve those outcomes.

It has been noted, however, that beyond participation in everyday work activities there are particular ways in which learning in and through work can be promoted, enhanced and augmented. That is, to make workplaces more effective learning environments includes addressing the limitations of learning in and through work. The sequencing of activities in which workers can engage, in ways that support their learning incrementally and embedded in work activities, is proposed here as being a long-standing but essential feature of organizing a pathway of learning experiences for both novices and experienced workers, referred to as the learning curriculum. Then, there are practices in the workplace that are pedagogic insofar as they actively support learning, some of which are enacted by more experienced workers and others arising from interaction with artifacts and opportunities presented by work activities, which are inherently aligned with effective learning. Then, the basis by which workers come to actively engage with the workplace activities and interactions they are afforded emphasizes the importance of their role in learning and development through and across working life. Allied to that engagement are sets of personal epistemological practices that these workers can enact to promote their learning. So, much of what is proposed here draws upon earlier accounts and approaches to promote learning through work and bringing about workplace change, but the nuances and means of achieving particular outcomes are informed by more recent accounts of how learning can be supported in and through workplaces. For instance, the need to engage with and also to communicate through digital technologies is now adding both specific goals and processes associated with learning in and through work. These emerging trends, again, emphasize the importance of having the means by which the knowledge to be learnt can be made accessible to workers and that they have the interests and capacities to engage with and learn the knowledge effectively.

8.7. References

Ainley, P. and Rainbird, H. (eds) (1999). *Apprenticeship: Towards a New Paradigm of Learning*. Kogan Page, London.

Anvik, C., Vedeler, J.S., Wegener, C., Slettebø, Å., Ødegård, A. (2020). Practice-based learning and innovation in nursing homes. *Journal of Workplace Learning*, 32(2), 122–134.

Avby, G. (2016). Organizing for deliberate practice through workplace reflection. In *Supporting Learning Across Working Life: Models, Processes and Practices*, Billett, S., Dymock, D., Choy, S. (eds). Springer, Dordrecht.

Bakon, S., Wirihana, L., Christensen, M., Craft, J. (2017). Nursing handovers: An integrative review of the different models and processes available. *International Journal of Nursing Practice*, 23(2).

Baldwin, J.M. (1898). On selective thinking. *The Psychological Review*, V(1), 1–24.

Berryman, S. (1993). Learning for the workplace. *Review of Research in Education*, 19, 343–401.

Billett, S. (2001). *Learning in the Workplace: Strategies for Effective Practice*. Allen and Unwin, Sydney.

Billett, S. (2006). Constituting the workplace curriculum. *Journal of Curriculum Studies*, 38(1), 31–48.

Billett, S. (2009). Personal epistemologies, work and learning. *Educational Research Review*, 4(3), 210–219.

Billett, S. (2014). *Mimetic Learning at Work: Learning in the Circumstances of Practice*. Springer, Dordrecht.

Billett, S. (2015). Work, discretion and learning: Processes of learning and development at work. *International Journal of Training Research*, 13(3), 214–230.

Billett, S. (2016). Apprenticeship as a mode of learning and model of education. *Education & Training*, 58(6), 613–628.

Billett, S. (2021). Mediating worklife learning and the digitalization of work. *British Journal of Educational Technology*, 52(4), 1580–1593. doi: 10.1111/1467-8535.13115.

Billett, S. and Newton, J. (2010). A learning practice. Conceptualising professional lifelong learning for the health-care sector. In *Beyond Reflective Practice: New Approaches to Professional Lifelong Learning*, Bradbury, H., Frost, N., Kilminster, S., Zukas, M. (eds). Routledge, Abingdon-on-Thames.

Billett, S. and Noble, C. (2020). Utilizing pedagogically rich work activities to promote professional learning. *Éducation & didactique*, 14(3), 137–150.

Billett, S., Noble, C., Sweet, L. (2017). Pedagogically-rich activities in hospital work: Handovers, ward rounds and team meetings. In *A Practical Guide for Learning and Teaching in a Clinical Context*, Delany, C. and Molloy, L. (eds). Elsevier, Melbourne.

Billett, S., Yang, S., Chai, A., Fang Tai, J., Lee, M., Alhadad, S. (2018). The co-occurrence of learning through work and innovations at work: Prospect for personal, workplace and cultural change. In *9th International Conference EARLI SIG 14 Learning and Professional Development*. University of Geneva, Geneva.

Billett, S., Yang, S., Chai, A., Tai, J.F., Lee, M., Alhadad, S. (2021). Remaking and transforming cultural practices: Exploring the co-occurrence of work, learning and innovation. In *Creativity and Learning: Contexts, Processes and Impact*, Collin, K., Glaveanu, V., Lemmetty, S., Forsman, P. (eds). Palgrave Macmillan, London.

Bunn, S. (1999). The nomad's apprentice: Different kinds of apprenticeship among Kyrgyz nomads in Central Asia. In *Apprenticeship: Towards a New Paradigm of Learning*, Ainley, P. and Rainbird, H. (eds). Kogan Page, London.

Chan, S. (2013). Learning through apprenticeship: Belonging to a workplace, becoming and being. *Vocations and Learning*, 6(3), 367–383.

Cooper, L., Orrel, J., Bowden, M. (2010). *Work Integrated Learning: A Guide to Effective Practice*. Routledge, London.

Dean, B.A. and Sykes, C. (2021). How students learn on placement: Transitioning placement practices in work-integrated learning. *Vocations and Learning*, 14(1), 147–164.

Deissinger, T. and Hellwig, S. (2005). Apprenticeships in Germany: Modernising the dual system. *Education and Training*, 47(4/5), 312–324.

Deissinger, T. and Ott, M. (2016). Tertiarisation of vocational education and training and its implications – Problems and issues in Germany and France. In *Education Policy: Mapping the Landscape of Scope*, Bohlinger, S., Dang, T.K.A., Glatt, M. (eds). Peter Laing, Frankfurt.

Dornan, T. (2012). Workplace learning. *Perspectives on Medical Education*, 1(1), 15–23.

Ellström, P.E. (2001). Integrating learning and work: Problems and prospects. *Human Resource Development Quarterly*, 12(4), 421–435.

Eppich, W., Rethans, J.J., Tueunissen, P.W., Dornan, T. (2016). Learning to work together through talk: Continuing professional development in medicine. In *Supporting Learning Across Working Life: Models, Processes and Practices*, Billett, S., Dymock, D., Choy, S. (eds). Springer, Dordrecht.

Epstein, S.R. (2005). Transferring technical knowledge and innovating in Europe 1200–1800. Working paper [Online]. Available at: http://eprints.lse.ac.uk/22547/1/0105Epstein.pdf.

Ericsson, K.A. (2006). The influence of experience and deliberate practice on the development of superior expert performance. In *The Cambridge Handbook of Expertise and Expert Performance*, Ericsson, K.A., Charness, N., Feltowich, P.J., Hoffmann, R.R. (eds). Cambridge University Press, Cambridge.

Esmond, B. (2020). Emerging apprenticeship practitioner roles in England: Conceptualising the subaltern educator. *Vocations and Learning*, 13(2), 179–196.

Fodor, J.A. (1983). *The Modularity of Mind.* MIT Press, Cambridge, MA.

Frost, N., Zukas, M., Bradbury, H., Kilminster, S. (eds) (2010). *Beyond Reflective Practice: New Approaches to Professional Lifelong Learning.* Routledge, London.

von Glasersfeld, E. (1987). Learning as a constructive activity. In *Problems of Representation in the Teaching and Learning of Mathematics*, Janvier, C. (ed.). Lawrence Erlbaum, Hillsdale.

Goller, M., Steffen, B., Harteis, C. (2019). Becoming a nurse aide: An investigation of an existing workplace curriculum in a nursing home. *Vocations and Learning*, 12, 67–85. doi: 10.1007/s12186-018-9209-z.

Gonon, P. (2009). 'Efficiency' and 'vocationalism' as structuring principles of industrial education in the USA. *Vocations and Learning: Studies in Vocational and Professional Education*, 2(2), 75–86.

Gowlland, G. (2012). Learning craft skills in China: Apprenticeship and social capital in an artisan community of practice. *Anthropology and Education Quarterly*, 43(4), 358–371.

Greinert, W.-D. (2002). Towards a history of vocational education and training (VET) in Europe in a comparative perspective. In *Proceedings of the First International Conference*, volume II. Cedefop, Florence.

Halvarsson Lundkvist, A. and Gustavsson, M. (2018). Conditions for employee learning and innovation – Interweaving competence development activities provided by a workplace development programme with everyday work activities in SMEs. *Vocations and Learning*, 11(1), 45–63. doi: 10.1007/s12186-017-9179-6.

Harteis, C. (2018a). Supporting learning at work in an era of digitalization of work. In *Work-based Learning as a Pathway to Competence-based Education*, Bahl, A. and Dietzen, A. (eds). Barbara Budrich, Opladen.

Harteis, C. (ed.) (2018b). *The Impact of Digitalization in the Workplace*, volume 21. Springer, Dordrecht.

Henderson, A. and Alexander, H. (2011). Maximising the integration of medical and nursing students in clinical learning environments: An Australian perspective. In *Developing Learning Professionals: Integrating Experiences in University and Practice Settings*, Billett, S. and Henderson, A. (eds). Springer, Dordrecht.

Hoyrup, S., Bonnafous-Boucher, M., Hasse, C., Lotz, M., Molller, K. (eds) (2012). *Employee-Driven Innovation: A New Approach*. Palgrave-Macmillan, Basingstoke.

International Labour Organisation (2015). *Upgrading Informal Apprenticeships: A Bottom Up Approach*. ILO, Beriut.

Jordan, B. (2011). *The Double Helix of Learning: Knowledge Transfer in Traditional and Techno-centric Communities*. Palo Alto Research Center, Palo Alto.

Kothari, A. and Wathen, C.N. (2017). Integrated knowledge translation: Digging deeper, moving forward. *Journal of Epidemiology and Community Health*, 71(6), 619–623.

Lancy, D.F. (2012). "First you must master pain": The nature and purpose of apprenticeship. *Anthropology of Work Review*, 33(2), 113–126.

Lave, J. (1990). The culture of acquisition and the practice of understanding. In *Cultural Psychology*, Stigler, J.W., Shweder, R.A., Herdt, G. (eds). Cambridge University Press, Cambridge.

Lave, J., Murtaugh, M., de la Roche, O. (1984). The dialectic of arithmetic in grocery shopping. In *Everyday Cognition: Its Development in Social Context*, Rogoff, B. and Lave, J. (eds). Harvard University Press, Cambridge, MA.

Lui, K.W.R. and Choy, S. (2020). How ethnic cultures influence practice-based learning. *Journal of Workplace Learning*, 32(3), 183–197.

Makovichy, N. (2010). 'Something to talk about': Notation and knowledge-making among Central Slovak lace-makers. *Journal of the Royal Anthropological Institute (NS)*, 16(Supplement S1), 80–99.

Malle, B.F., Moses, L.J., Baldwin, D.A. (2001). Introduction: The significance of intentionality. In *Intentions and Intentionality: Foundations of Social Cognition*, Malle, B.F., Moses, L.J., Baldwin, D.A. (eds). The MIT Press, Cambridge, MA.

Marchand, T.H.J. (2008). Muscles, morals and mind: Craft apprenticeship and the formation of person. *British Journal of Education Studies*, 56(3), 245–271.

Marsick, V.J. (1988). Learning in the workplace: The case for reflectivity and critical reflectivity. *Adult Education Quarterly*, 38(4), 187–198.

Marsick, V.J. and Watkins, K. (1990). *Informal and Incidental Learning in the Workplace*. Routledge, London.

Mishler, E.G. (2004). *Storylines: Craftartists' Narratives of Identity*. Harvard University Press, Cambridge, MA.

Moore, A.L. and Klein, J.D. (2020). Facilitating informal learning at work. *TechTrends*, 64(2), 219–228.

Noble, C., Brazil, V., Teasdale, T., Forbes, M., Billett, S. (2017). Developing junior doctors' prescribing practices through collaborative practice: Sustaining and transforming the practice of communities. *Journal of Interprofessional Care*, 31(2), 263–272. doi: 10.1080/13561820.2016.1254164.

Noble, C., Billett, S., Hilder, J., Teodorczuk, A., Ajjawi, R. (2019). Enriching medical trainees' learning through practice: A video reflexive ethnography study protocol. *BMJ Open*, 9(8).

Organisation of Economic Co-operation and Development (1996). *Lifelong Learning for All.* OECD, Paris.

Organisation for Economic Co-operation and Development (2006). *Live Longer, Work Longer*. OECD, Paris.

Organisation for Economic Co-operation and Development (2013). *OECD Skills Outlook 2013: First Results from the Survey of Adult Skills.* OECD, Paris.

Pelissier, C. (1991). The anthropology of teaching and learning. *Annual Review of Anthropology*, 20, 75–95.

Piaget, J. (1971). *Structuralism* (translated and edited by C. Maschler). Routledge & Kegan Paul, London.

Pylväs, L., Nokelainen, P., Rintala, H. (2018). Finnish apprenticeship training stakeholders' perceptions of vocational expertise and experiences of workplace learning and guidance. *Vocations and Learning*, 11(2), 223–243.

Raizen, S.A. (1991). *Learning and Work: The Research Base. Vocational Education and Training for Youth: Towards Coherent Policy and Practice*. OECD, Paris.

Rice, T. (2010). Learning to listen: Auscultation and the transmission of auditory knowledge. *Journal of the Royal Anthropological Institute (NS)*, 16, S41–S61.

Rintala, H. and Nokelainen, P. (2019). Vocational education and learners' experienced workplace curriculum. *Vocations and Learning*, 1–18.

Rogoff, B. (1990). *Apprenticeship in Thinking – Cognitive Development in Social Context.* Oxford University Press, New York.

Rogoff, B. and Lave, J. (eds) (1984). *Everyday Cognition: Its Development in Social Context.* Harvard University Press, Cambridge, MA.

Roodhouse, S. (2007). Special issue introduction. *Education + Training*, 49(3), 161–169.

Rose, N. (1990). *Governing the Soul: The Shaping of the Private Self.* Routledge, London.

Sennett, R. (2008). *The Craftsman.* Yale University Press, New Haven/London.

Singleton, J. (1989). The Japanese folkcraft pottery apprenticeship: Cultural patterns of an educational institution. In *Apprenticeship: From Theory to Method and Back Again*, Coy, M.W. (ed.). SUNY, New York.

Tobias, S. (1994). Interest, prior knowledge, and learning. *Review of Educational Research*, 64(1), 37–54.

Tomasello, M. (2004). Learning through others. *Daedalus*, 133(1), 51–58.

Valsiner, J. (1998). *The Guided Mind: A Sociogenetic Approach to Personality*. Harvard University Press, Cambridge, MA.

Vygotsky, L.S. (1978). *Mind in Society – The Development of Higher Psychological Processes*. Harvard University Press, Cambridge, MA.

Wallin, A., Pylväs, L., Nokelainen, P. (2020). Government workers' stories about professional development in a digitalized working life. *Vocations and Learning*, 13(3), 439–458.

Webb, E. (1999). Making meaning: Language for learning. In *Apprenticeship: Towards a New Paradigm of Learning*, Ainley, P. and Rainbird, H. (eds). Kogan Page, London.

Wertsch, J. and Tulviste, P. (1992). L.S. Vygotsky and contemporary developmental psychology. *Developmental Psychology*, 28(4), 548–557.

9

The Activity Analysis Approach in Education and Training Sciences: Challenges, Principles and Perspectives

Joris THIEVENAZ
LIRTES, Université Paris-Est Créteil, France

Education and training sciences (ETS) share the particularity of taking phenomena related to socioprofessional practices (teaching, training, tutoring, advice, etc.) as their research object, with other scientific disciplines. Far from being confined to the study of learning and knowledge appropriation at school or in the academic field, the study object of research in this discipline also comprises processes and situations which contribute to the *subject's training*[1], that is to say, to the being's maturation dynamics in a plurality of contexts and this, throughout a lifetime.

It is from this scientific perspective that favors the analysis of human activities, both in their singular and situated dimension, that several research trends have emerged in France[2]. While the expression "*entry by activity*" has been addressed in a

1 The expression *subject training* is not to be understood in the sole sense of (knowledge, skills) acquisition with a didactic or pedagogic intention, but more broadly, as a construction dynamic of the person in its various capabilities (cognitive, emotional, social identity, etc.).

2 This chapter tackles the analysis of on-the-job human activities based on various scientific approaches developed in France since the 1990s. The reader that is interested in other concepts and methods of work analysis developed in the English-speaking world may refer to the contribution of S. Billet in this work.

Encyclopedia of Professionalization,
coordinated by Didier DEMAZIÈRE and Richard WITTORSKI.

number of recent studies and tends to be valorized by the actors who employ it, the use of this term underlies very different scientific approaches. Frequently used as the discipline which studies the relationships between work, learning and development, it contributes in exploring the subject's professionalization processes as the action is being carried out (Wittorski 2005), with reference to concepts and methods developed by different trends in the analysis of *work* and/or the *activity*[3].

This chapter aims to show synthetically to what extent these various research trends that share an interest in the activity analysis of subjects make it possible to document the processes which participate in their formation and/or that of professional groups at the workplace. The first part will review the scientific and praxeological challenges underlying different types of work. A second part will offer a condensed overview of the various theoretical trends. A third and final part will highlight how on-the-job activity analysis amply participates in a scientific perspective in human and social sciences (HSS), which can be qualified as *micrological.*

9.1. The intelligibility of relationships between work and learning: a scientific and praxeological challenge

Understanding the dynamics that contribute to the subject's professionalization *by*, *in* or *as a result of* work situations constitutes both a scientific and praxeological challenge. For the HSS, the question is to better understand the transformations that take place within the subject, in the flow of action and during the accomplishment of tasks oriented by production purposes. For the professional environment, the challenge is to encourage what is sometimes called *the rise in skills* or the *development of the actors' power to act* through better knowledge of these phenomena. Whether from an intelligibility or intervention perspective, the study of professional situations is, in all cases, underpinned by strong challenges for the researcher and for the participant (trainer, designer, tutor, caregiver, accompanist, etc.).

9.1.1. *A field of research constituting education and training sciences*

Based on the observation that professional situations, accomplished through production challenges (of goods and/or services) are also spaces for self-construction, knowledge acquisition and even for the emancipation of those who experience them, numerous research studies in the field explore the formative

3 The author would like to thank B. Albero, S. Eloi, L. Filliettaz, J. Guérin, A. Mouchet, P. Olry and R. Wittorski.

potential of labor. While the problem of human development as it acts on the world is a classic concern in HSS, the analysis of work in relation to education and training questions is a relatively more recent area of study. In the French-speaking world, several research trends have developed since the 1990s, whose theoretical frameworks and methods have been supported by contributions from different disciplines relating to work sciences (ergonomics, work psychology, work sociology, etc.). They share the same postulate: by the very fact of its fulfillment and apart from any formative or empowering intention, work is potentially formative for those who experience it. Apart from the interest in training schemes (Albero 2010), which are more or less formalized and/or integrated into the professional space, and the more or less voluntary transmission phenomena that take place there, the question is to better understand to what extent professionals enrich their experience and are transformed by the very fact of their work activity.

It is from this perspective that numerous works in the discipline have been carried out for over 30 years in a wide diversity of professional environments to identify and make certain phenomena more intelligible, such as "experience-based knowledge|, "on-the-job learning" or even "empowering environments" (see Chapter 7). While other disciplines take professional activities as their study object, research in education and, more specifically, training studies the connections between work and learning in their relationship to formative concerns: "the question here will be to identify and characterize what work analysis can 'produce', which can be used in the field of training" (Savoyant 2006, p. 127). The aim of work analysis is less to model the entirety of the dynamics which contribute to the production system than to succeed in identifying development invariants on-the-job.

It is by postulating that a large part of learning that participates in the continuous reconstruction of experience emerges in the flow of activity and in the multiple folds of professional action that numerous research projects in the field of education and training have been carried out since the end of the 1980s. Despite the abundant work carried out in this perspective, this area of inquiry remains open, as the phenomena pertaining to the subject's subjectivity remain largely unknown from the point of view of their origin, their own dynamics and their impacts on the subject. A difficulty inherent in this type of approach stems from the fact that the métier observed is often learned "on-the-job", but also often without the knowledge of those who experience it. A paradox the researcher faces is related to no one being more competent than the professional to speak about their métier and to testify to their experiences. However, no one is in more difficulty than the professional in the exercise of their activity to explain the goals, means and resources which participate in their action, as these processes seem incorporated and intimately linked to the conduct of their action.

When actors are aware of learning something while carrying out their action, they often find it challenging to describe what they learn and how they learn it. The action *on* and *in* the world is a source of profound learning every day, but the words to express it are often lacking. It is starting from this observation that a number of methods have been developed with the goal of enabling actors to put their experience into words, in an attempt to approach a less visible dimension of work which falls within the domain of "constructions of meaning a subject operates as a result of and over its own activity" (Barbier 2018). These processes relating to the subject's mental, bodily and emotional life must always be understood and conceptualized. Apart from its interest in terms of knowledge production, this is a central and strategic challenge for those involved in education and training.

9.1.2. *A professional and social challenge*

Many actors in the fields of work and training (trainers, teachers, managers, designers, guides, caregivers, tutors, *coaches*, advisors, etc.) are confronted daily with the challenge of better understanding, identifying and acting on the drivers of professional development. Knowledge of the activities actually carried out by the subjects when performing their daily tasks – not to mention the conditions in which these interventions can prove to be learning or formative – constitutes a major challenge for research. This means not being content with a superficial description of the professional activity, but on the contrary, studying the principles that organize it and the active properties of situations as closely as possible:

> The trainer needs to know in detail what allows the worker to perform the job. This analysis cannot be satisfied with approximation, because it is intended for action. It has to focus on describing the work's content. From the outset, the trainer must strive to obtain information about the work which will later facilitate the transfer from the learning situation to the real work situation (de Montmollin 1974, p. 41).

Work analysis equips the activity of those involved by providing instruments (concepts and methods), which document and guide the conduct of the supporting or training action by making it more intelligible. Depending on the contexts and socioeconomic challenges, this may involve providing benchmarks concerning the goals pursued by the actors on a daily basis, the tools and means they develop, the difficulties and obstacles they come across, the risks they take and the types of learning and innovations that they sometimes deploy without their knowing. Work analysis is not only the researcher's prerogative, but also a major challenge for the trainer who, for the purposes of intervention, cannot be content to rely solely on legislation, prescriptive documents or benchmarks developed a priori or at a distance from the profession which, while being structuring, still runs the risk of formalizing lists of knowledge, tasks or know-how decoupled from the actual challenges of professional exercise.

The careful study of the work activity and the forms of learning related to it support the intervention of the trainer in various ways through (1) the design of tools, programs, support tools and educational paths supported by knowledge of real work, which is not limited to what is prescribed or displayed; (2) identification of what constitutes the core of the métier it trains for; (3) information on what, in a given class of professional situations, constitutes benchmarks or invariant development principles; (4) understanding of the constraints weighing on the actors; (5) the identification of resources and professional knowledge invoked by actors (experts or novices) to act effectively in the situations they encounter; and (6) knowledge of the types of problems to be solved and the intellectual processes to be implemented to achieve them.

It is easy to understand the convergence of scientific and praxeological interests, for the analysis of work in the field of education and training leading to the production of knowledge that is both heuristic and scientifically valid, potentially usable by professionals in the field, as resources for training and subject accompaniment.

Even if today it is commonly accepted that all professionals benefit from becoming familiar with the concepts, methods and knowledge resulting from work analysis, this has not always been the case. Studying the relationships between work, learning and training supposes the meeting between two fields, which have been historically constituted in a relatively separate way.

9.1.3. *A rapprochement between two worlds: work and professional training*

The analysis of work in the field of education and training is a research and intervention perspective, which finds its origin in the rapprochement between two types of cultures: work sciences and adult training. Even if this dynamic is currently well known, history shows that such an encounter is relatively recent, since the work environment and professional training constituted themselves as relatively autonomous sectors, at a distance from one another[4]. It is for this reason that Guy

4 Apart from the founding works of French-language ergonomics, which will be discussed later in the text.

Jobert speaks of an "unfortunate relationship" (1993) between trainers and the world of work. The world of adult training has kept a relatively long distance from the actual activity of subjects on-the-job, not only due to cultural and identity reasons, but also relating to organizational habits and the feasibility of such a connection: "Adult training has missed out on the contributions of work psychology, and this for many years" (Pastré 2011, p. 27).

It was therefore necessary to bring together these two worlds of work analysis and professional training, so that several research and intervention perspectives could be initiated. Today, these can be described as *activity analysis approaches in the field of education and training*. The meeting of the world of adult training with the principles, concepts and methods of work analysis was encouraged, at the end of the 20th century, by a cultural, economic and political context marked by profound changes in work and the new employability challenges that resulted from it. This translates the emergence of different trends favoring an approach through the analysis of activity and/or work[5].

9.2. The reference in education and training sciences to different theoretical trends or approaches which study the way in which humans act, learn and transform themselves by producing

Far from constituting a homogeneous trend or a unified research perspective, the activity analysis approach brings together diverse research orientations that have the primacy given to the singularity of human activity in common. Insofar as several works and synthesis articles have already presented each of these trends, orientations or approaches[6] in detail, this chapter aims to clarify what constitutes their specificities and convergences. Resuming a first synthetic perspective (Albero and Guérin 2014), this chapter aims to highlight the shared postulates and unifying principles, which guide or structure such approaches, and support research work in the discipline. Let us briefly mention the concepts and methods they are based upon.

5 It is interesting to observe that the terms "activity analysis" and "work analysis" are preferred, depending on the approach, to that of "practice analysis" which is too imbued with the language of intervention in which ambiguity persists as to whether it is an analysis *on* the practice, *in* the practice itself by the professionals who carry it out, *by* practice for changes, etc. The syntagm "theories of activity" is an extension of pioneering work (notably that of Soviet psychology) to which most authors refer.

6 See in particular on this subject: Marcel et al. 2002; Champy-Remoussenard 2005; Durand 2009; Albero and Guérin 2014.

9.2.1. *Pioneering work establishing the principles of work analysis in relation to formative and developmental concerns*

Among the pioneering works which evoke the principles of work analysis in relation to formative concerns, the founding work of psychologists Ombredan and Faverge (1955) entitled "*L'analyse du travail*" brings about a turning point in the way we observe and study professional activities. In fact, they broke with the then usual psycho-technical approaches (centered on the principle of measurement and effectiveness) and favored a more comprehensive approach, in which "it is the attitude at work that is analyzed" (ibid., p. 31). Accounting for the difficulties encountered by *the agent*, collecting traces of its activity or observing established forms of communication, is part of a desire to "develop a training program" (ibid., p. 222). Following up on this work, several publications by de Montmollin (1974) established the principles of an "work analysis prior to training". Here again, a paradigm shift operates, with the cancellation of a strictly evaluative viewpoint of human performance at work and the increase of a comprehensive analysis of the gestures, attitudes and modes of reasoning that organize the activity, as well as the ways in which everyone learns to hold their work position: "the work study methods recommended here are far more concerned with learning tasks than the old methods, which mainly studied an already trained worker" (ibid. 1961, p. 23). While it is not possible, in the context of this chapter, to return in detail to all the pioneering work founding the approaches presented – in particular those developed within the framework of the Moscow Institute of Psychology – it is nonetheless important to remember that in the field of education, a proposal for generic modeling of human activity has been carried out (Linard 1989, 2001), supported in particular by the work of Leontiev and Greimas, to provide "a common organizing framework for understanding the activity in general, formalizing learning and designing systems with training purposes" (ibid. 2001, p. 217). In the wake of this founding work, different research streams have been developed since the beginning of the 1990s.

9.2.2. *Professional didactics*

Defined by P. Pastré in 1992 as "work analysis for training", this approach aims to explicitly account for the cognitive models and organizing principles of the subject's activity (operational model according to Pastré 2011), with a didactic and formative aim. This research trend is based on at least three types of theoretical contributions: (1) those of Soviet psychology (Galperine 1966; Ochanine 1969; Leontiev 1972); (2) those of French-language ergonomics or psycho-ergonomics

(Ombredane and Faverge 1955; de Montmollin 1974; Rabardel 1995; Leplat 1997); (3) those of the genetic psychology of Piaget (1967, 1974) and disciplinary didactics (Vergnaud 1990; Brousseau 1998). By taking up the concept of "scheme" borrowed from Jean Piaget and developed more recently by Gérard Vergnaud, various concepts specifically adapted to fieldwork were developed to account for the activity's cognitive dimension in this context. This work is based on the hypothesis according to which it is only possible to design training tools, situations and environments, as close as possible to so-called real work, based on knowledge of the variety and the diversity of models that guide the action of several professionals and the principles that generate their activity. The first research conducted mainly focused on industrial production (Pastré 1992) and agriculture (Caens-Martin 1999), whereas other works focused on the areas of service and human interactions, documenting this approach (Mayen 1997; Olry and Bautzer 2003; Piot 2006). It was based on the following postulate: training professionals to exercise their métier presupposes understanding the operational models, which are developed and transformed in and through the activity. The term *didactic* stresses the intention of transforming a reference work situation into a training situation. Various tools, orientations and training courses were thus developed, inspired in the results of this research (simulator, career path, situation benchmarks, etc.).

9.2.3. *The activity clinic and the psycho-dynamics of work*

Based on the dual heritage of French-language ergonomics and work psychopathology, and drawing on the work of Vygotski (1934) and Bakhtin (1975), the activity clinic is part of a historico-cultural approach to the subject's development. This theoretical trend is based on the distinction between real activity and the reality of activity: "What has been achieved does not have a monopoly on the real" (Clot 2011, p.17). Analyzing and understanding work involves taking into account, not only what the subject accomplishes while conducting an action, but also the hindered activity (Clot 1999), what the subject cannot or does not know how to do, what they try to do unsuccessfully, the obstacles encountered, etc. This hindered activity can, in certain cases, constitute an essential dimension of human work. The term *clinic* is used for defining an approach that aims not only to understand, but also to "act and transform the profession". In this approach, the epistemic side of research is inseparable from its praxeological side. It renews the structuring distinction between task and activity (see below), at the origin of methods frequently used in research in education and training: instructions to the replacement[7] (Oddone et al. 1977 [1981]) and simple and/or cross self-confrontation[8] (Clot 1999, 2008). Without going into the detail of their process, it is possible to observe that they are methodologically based on the use of language, dialogue and controversy. This focus on language activity and the developmental potential attached to it is an opportunity to introduce two other structuring concepts of this approach (drawn from Bakhtin's work): the professional genre[9] and style[10] (Clot and Faïta 2000). It is also based on the observation that "grasping the dynamic between style and genre relationships" (ibid. p. 18) is a source of pathogenic situations and an amputation of the power to act. The activity clinic makes suffering at work an object of investigation.

Thanks to the contributions of work psychopathology, psychoanalysis and the founding distinction between task and activity proposed by ergonomists, work psycho-dynamics (Dejours 1980) explores the activity's subjective and intersubjective dimension, postulating that "the problem of suffering arises when there is no room left for the metaphorical play of desire in the design activity" (ibid. 2016, p. 45). In other words, when humans are no longer able to make a contribution and to invoke intelligence for performing the task entrusted to them, suffering arises. While the psycho-dynamics of work falls under psychology as a discipline, many researchers in education and training regularly use work to question the notion of creativity in professional action.

9.2.4. *Ergology*

Inspired in the works of Georges Canguilhem and Karl Marx, ergology also takes as a working perspective the intelligibility of the way in which the subject, in order to act, is called upon to reinterpret, reformulate or reconstruct the task

---

7 This method, developed by I. Oddone in the 1970s, aims to understand "the way in which workers develop experience in the factory and structure their behavior" (p. 55). It is based on the formulation of the following instructions addressed to workers who volunteered to participate in the study: "Suppose that I am your replacement and that tomorrow I find myself in a situation of having to replace you at work. What are the instructions you would like to transmit to me so that no one notices the substitution?".

8 "The self-confrontation situation is one where the operators, exposed to the image of their own work, first put into words, for the use of the partner-spectator, what they think are the constants" (Clot and Faïta 2000, p. 26).

9 "What they know they must do thanks to a community of presupposed evaluations, without the need to re-specify the task each time it arises" (Clot and Faïta 2000, p. 11).

10 "Style can therefore be defined as a metamorphosis of the genre in action" (Clot and Faïta 2000, p. 15).

prescribed to them. To designate this phenomenon, ergonomists traditionally use the syntagm "task redefinition" (Leplat 1997), "renormalization" (Schwartz 1998) being conceptualized in this research trend. It emphasizes the fact that the work activity reorganizes and reformulates former standards in order to adjust them in a way adapted to the context and the person:

> To understand its drama, an ergological process implies considering it as far as possible as the singular moment of an indefinitely renewed debate between antecedent norms and attempts at renormalization [...] and which therefore requires retelling the story [...] of this life center seeking to be renormalized (living being, human individual, collective entities at various levels of coherence) (ibid. p. 23).

This research and intervention perspective in professional environments is part of a philosophical approach to work, and it reinforces numerous research works in the field (Felix and Saujat 2015).

9.2.5. *The course of action*

From the perspective of cognitive anthropology, the course of action (Theureau 1992) is based on a semiological activity analysis, inspired in the semiotics of Peirce (1935) and the theory of enaction developed by Maturana and Varela (1980). The central postulate of this approach is that humans think and act by interpreting signs[11]. According to this approach, the human being is constantly transformed by a coupling dynamic, which operates with the action's environment. Sequential activity analysis identifies which possibilities are actualized or not in the situation for the purposes of performing the action. From a methodological point of view, the activity's video-recording and the arrangement of first- and second-degree self-confrontation situations make it possible to identify what is signaling for the actor at each moment (called representation) and the elementary action (practical, symbolic, communication, etc.) that results from it. A specificity of this enactive approach of the experience is that it deconstructs the flow of individual and/or collective activity into elementary units, and then proceeds with their sequential modeling. With reference to Lakatos (1978/1994), this approach proposes three research programs: philosophical, empirical and technological (Pinsky 1992). Numerous areas have been studied since the beginning of the 1990s, especially the teaching of physical and sporting activities, the activity of students in primary school or middle school, or even the training of apprentice teachers.

11 "Each unit of action emerges as a unit of meaning, specified by the actor at each moment throughout its activity" (Durand 2009, p. 849).

9.2.6. *The psycho-phenomenological approach to experience*

The psycho-phenomenological approach to experience initiated by Pierre Vermersch (1994) is based on the observation that humans in action permanently memorize a set of data relating to their experience. This process that Edmund Husserl calls "retention" (1905/1964) is not immediately accessible by the subject, but constitutes a background while conducting an action. The whole challenge of the explanatory interview (Vermersch 1994) is that it supports the subject thanks to precise guidance enabling them to evoke the action experienced in a past, singular and specified situation and the information perceived and memorized on that occasion, even without being aware of it. This method aims to access the sensory information captured by the subject at a specific moment in their experience, but also to put into words the concrete progress of their activity and the mental and emotional processes, which accompany the experience of the action. This process involves an "evocation" of the subject, during which they verbalize the past activity in *relived* (or revival) mode by "a transition from non-consciousness to reflected consciousness in the aftermath" (ibid. 1994, p. 75). While it is not possible to describe in detail the language processes implemented in such a method, it is nevertheless interesting to note that, due to the experiential and formative potentialities it contains, it is regularly used as an intervention tool in education and training (Faingold 2011; Mouchet 2018).

9.2.7. *Linguistic and interactional approaches*

From the dual heritage of the psychology of Piaget (1967, 1974) and Vygostki (1934) and the contributions of linguistics, a research trend close to psycho-linguistics and/or language analysis emerged, developed at the University of Geneva, in the late 1980s. Two main methods of analysis can be differentiated. The first focuses on the way in which the use of language is likely to constitute an instrument of representation and description of action. Works then mobilize the instruments of discourse analysis to understand how language resources contribute to shape action and the way in which the use of language is likely to accomplish and coordinate actions in the world (Bronckart and Bulea 2006). A second orientation is interested in the pragmatic and active dimension of language, which enables actions to be performed and coordinated. These works borrow more from sociolinguistics and the micro-sociology of verbal interactions, with the goal of understanding how the participants of the interaction carry out an intersubjective interpretation process at the moment when they speak to one another, so as to act together (Filliettaz and Zogmal 2020). These two approaches, which are not antagonistic but complementary, therefore pursue distinct analytical goals in their way of analyzing activity (Filliettaz et al. 2021).

9.2.8. *The epistemology of activity analysis frameworks*

This perspective takes the analytical tools used by approaches centered on activity analysis as its study object. The materials explored by this second-level analysis are no longer the verbalizations of actors or the observations carried out in the field, but the matter which feeds such research. This work perspective can be qualified as an approach to the intelligibility of tools for analyzing human activities. This is likely to take several forms, depending on the theoretical references and postulates used. It may involve establishing a set of distinctions on the semantico-epistemological level for the research activity (Barbier 2000) in order to differentiate the thought frameworks of subjects and the epistemic frameworks of research about the subjects' activity. It may also involve studying to what extent the different orientations for analyzing activity are likely to be understood as an organizing framework and *integrator* research in HSS (Linard 1989; Albero and Brassac 2013; Albero and Guérin 2014). Another line of research would be to tackle an archeology of activity analysis categories (Thievenaz 2021) by studying the interpretations of different researchers on a given theme (according to different postulates, orientations or research cultures). The knowledge produced does not relate to the subjects' activity per se, but rather to the conceptual and analytical categories likely to be mobilized to account for it.

9.2.9. *A Deweyan approach to experience*

Other approaches or perspectives for analyzing human activities in work situations have emerged more recently. While there is no claim to founding a school or trend of analysis, the intention is to provide a complementary approach to the relationships between work and learning. It is from this perspective that a reading of experience on-the-job is done, inspired by the works of Dewey (1938 [2006]). Used as a framework for analyzing activity, it allows us to better understand the intellectual processes and forms of learning that occur when the subject comes across problematic, conflicting situations or situations giving rise to doubt in a work situation (Thievenaz 2012, 2014, 2019a, 2019b; Mayen 2013; Piot 2017).

This quick overview of the main trends which underpin research in ETS, in particular those which study human activity in a work situation, is far from exhaustive. Other research trends or perspectives could have been mentioned[12]. However, the aim of this chapter is less exhaustiveness than the interest of showing how these main approaches are used in research in the discipline, and how they share general principles on human activity in a working context, beyond their respective specificities.

9.3. Principles which structure and guide research work in education and training sciences

While each activity analysis approach has its specificities (epistemological point of view, theoretical framework, concepts, methods), it seems decisive for the discipline to note the invariant principles or shared conceptions that structure such work. This is less for the sake of homogenizing a landscape which is by definition plural than to highlight foundations that are sometimes forgotten because they are implicitly shared.

9.3.1. *The enigmatic nature of work and activities*

If there is an implicit principle shared by all of the approaches presented here, it is that work, and more specifically, human activity at work, is studied as part of an enigmatic process which is not directly disclosed, but requires being studied in detail to try to understand its mechanisms and principles. Numerous expressions have been used by the authors to reflect the fact that the goals pursued by the actors, the resources they mobilize, the obstacles they encounter and the learning they attain fall within the subjective sphere of action, and that these processes cannot be "grasped" in the literal sense of the term. They must regularly be inferred on the basis of behavioral, linguistic, postural and/or symbolic clues, which constitute the research material and therefore invite the greatest caution during its analysis.

9.3.2. *The distinction between task and activity*

Activity analysis approaches find their origin in a founding distinction introduced by ergonomists, according to which the task is not the activity. In other words, work that is *prescribed* is never equivalent to the work actually achieved. The action of a subject–agent–operator is never reduced to the execution of (prescribed) instructions, even if the organization gives them the status of executor.

12 In particular, that of the Joint action theory (Sensevy 2012), which is more of a didactic perspective in the field of teaching.

A large part of activity involves redefining the goals assigned to the subject and the means to achieve them, so as to gain in the capacity to adapt to the context and the means allocated, and in efficiency. Simply studying the prescriptions and instructions provided to agents does not grant access to the structuring dimension of human work, which also comprises activities for constructing meaning (Barbier 2000) based on action.

9.3.3. *The essential immersion in work spaces*

The different approaches emphasize the importance of studying human activities on-the-job as close as possible to the actual conditions in which they are carried out. Unlike other scientific approaches based on the principle of experimentation, "it is (…) no longer the laboratory which makes it possible to observe and interpret such activities, but 'the field' itself, in real life, and in real time" (de Montmollin 1986, p. 2). From this viewpoint, work analysis is ethnographically inspired, insofar as it involves sufficiently inserting ourselves into a professional environment/work collective to gather enough reliable information to organize it into scientifically valid data. This also means that, according to these approaches, discourse on practice or declarations about the métier are not sufficient. Understanding work also involves studying the actual processes that make it up.

9.3.4. *Taking into account both the productive and formative dimension of human activity*

By the very fact of its fulfillment, work is (potentially) an opportunity for development and maturation for those who perform it. Referring to the works of Marx[13] by most authors, this postulate is taken up by various precursors of Soviet psychology[14], then supporting the work of ergonomic psychology[15] and ETS. It is based on the principle that the activity of the situated subject should be studied from

13 "Work is, at first glance, an act that takes place between [humans] and nature. [Humans] themselves play the role of a natural power vis-à-vis nature. The forces with which the body is endowed, arms and legs, head and hands, the human sets things in motion, in order to assimilate materials by giving them a useful form for life. At the same time, the human acts through this movement on external nature and modifies it, modifies its own nature, and develops the faculties which lie dormant in it." (Marx 1867 [1969], p. 180).

14 Notably Leontiev (1972).

15 In particular, Rabardel (2005), who evokes the productive activity and the constructive activity to designate this phenomenon.

the viewpoint of its productive and developmental side, which research in the field explores professional action and its transformations.

9.3.5. *Taking into account the unique and situated character of human activities*

Studying the relationships between work and learning requires approaching the subject's activity, both in its singular and situated character. To understand the intellective, affective and operational processes implemented during the conduct of action, we have to approach the activity as a process that emerges in specific and concrete circumstances. To discuss this process, most authors refer to the work of Suchman (1985)[16] or Lave (1991 [2016]). More recently, other authors have proposed the notion of an ecological approach to learning and training situations (Albero et al. 2019) in order to emphasize the idea of taking into account human activities with close reference to the environment and the context where they evolve.

9.3.6. *An intention that is both epistemic and praxeological in the production of knowledge*

The analysis of work or activity on-the-job is guided by a double goal: producing scientific knowledge on human activity and the processes which participate in its development, and acting and/or participating in this development. In this view, all of the approaches or works presented in this chapter are to be positioned on "a structuring tension between episteme and praxis" (Albero 2010). This research has a dual purpose: understanding and transforming (Thievenaz et al. 2020). These two orientations demonstrate a desire to produce valid and scientifically supported knowledge, while participating in the development of the subject and work groups. The use of methods (self-confrontation, instructions to the replacement, interview to help with explanation, etc.), which constitute both research and intervention tools[17], is largely explained by this explicit double goal of research.

---

16 "To designate the alternative that ethnomethodology suggests—more a reformulation of the problem of purposeful action, and a research program, than an accomplished theory—I use the term situated action. That term underscores the fact that the course of action depends in essential ways upon the action's circumstances" (Suchman 1985, p. 35).

17 The fact of asking the subject to explain their activity or to comment on it, based on its video recording, potentially has a formative influence, whether intentional or not.

9.3.7. *The co-construction of valid knowledge*

In the field of work analysis, the expression "co-construction of knowledge" is regularly used to emphasize the cooperation between researchers and professionals in the investigation process. They are considered partners in the conduct of research, being recognized for their expertise, their capacity for analysis and reflexivity with regard to their own activity and the professional environment. For its part, the researcher brings theoretical and investigative expertise, putting (epistemological, conceptual, methodological, ethical, etc.) know-how at the service of the action's intelligibility.

9.3.8. *A holistic approach to human action*

Finally, it is interesting to observe that taking human activity as an object of study or a privileged entry into research work leads to producing knowledge, which re-examines the classic dichotomies that traditionally structure representations and actions. Understanding how the subject acts and is transformed by acting amounts to going beyond conceptions firmly anchored in reference models arranged around binary oppositions, which are as erroneous as they are lasting (theory/practice, reason/emotion, body/mind, work/leisure, etc.). The study of work and the way in which humans act confronts the challenge of resorting to a holistic approach in which the multiple dimensions of action are consubstantially intertwined in the flow of activity.

While it is not possible, in the context of this chapter, to address all the postulates or principles having gained consensus, it nonetheless seems useful to specify how the intelligibility of human activities in a work situation relates to an approach which can be described as micrological.

9.4. A micrological mode observation and analysis of human activities in a situation of production of goods and/or services

The approach that analyzes human activities in a production situation is part of a mode of knowledge that can be described as micrological (Laplantine 2003), when it concerns the activity's details – moreover, the details considered minor or even neglected – even if they prove decisive in the conduct of action. The understanding of the relationship between work and learning is then conditioned by the study of gestures, processes or phenomena that may appear at first sight to be ordinary, minor or trivial, and yet are decisive in the subject's activity and training.

9.4.1. *Attention to detail, the silent, the ordinary and the "almost nothing"*

It is by understanding the very small gestures, the discreet modulations, the minute, barely perceptible changes that it is possible to grasp how humans act and learn by acting. Such a research perspective is based on an observation: the subject's training does not only involve exceptional, visible or striking moments, but also discreet occurrences that are slightly glimpsed and yet decisive for those who experience them. To demonstrate the importance given to the small, the discreet, as well as to the tiny processes which contribute to the subject's activity, the authors explain that even the "smallest detail (…) can prove to be a capital element" (de Montmollin 1961, p. 35) and encourage us to "take into account the infinitesimal" (Schwartz 2012, p. 814). Conceptually, works regularly refer to notions such as that of "microgenesis" (Rabardel 1995; Pastré 2011) or the "micro-description of experience" (Vermercsh 2016, p. 566). In all cases, we have to reaffirm the importance of the infinitesimal and the tiny transformations that take place (continuously or not) in activity. This orientation opens the way to a better understanding of "situated micro-discoveries" (Thievenaz 2019b), which occur in the activity and progressively participate in the construction of the experience. This is how the development of the subject can be understood, in the infinitely small professional gestures and the transformations that accompany them. Beyond conceptual and methodological choices, we have to question this type of approach in terms of the epistemological questions it raises. A reflection should be carried out on the type of analysis scales adopted in the conduct of research and the degree of generalization of the knowledge produced.

9.4.2. *From the singular to the universal*

Observing and analyzing human activity in a production situation from the viewpoint of the smallest components is implicitly part of the postulate according to which "in every experience, we find the overall underlying qualitative whole which corresponds to the total organization of human activities which constitutes the mysterious human framework, and manifests it" (Dewey 1934 [2010], p. 325). Research favoring a micrological approach to activity and experience starts from the principle that the study of professional gestures in what they are most minimal, singular and situated, opens the way to a more general knowledge on human functioning. The identification and conceptualization of tiny processes implemented by a subject when carrying out an action in a specific context are understood as a possibility for producing knowledge, which can be generalized to other subjects and other types of situations. The conditions for a transition from situated observation to a generalized analysis are not always explained. One challenge is to show how (and to what extent) it is possible to find "the human universal in the individual" (Habermas 2021, p. 58).

9.5. Conclusion

This chapter presented the way in which the study of the subject's professionalization dynamics *by* and *in* its activity is based on a set of trends, approaches, methods and postulates, which document and structure the research activities of the field. To do this, it recalled the environment in which interest in the activity, from the viewpoint of adult training, emerged in France after the end of the 1980s, while presenting in a synthetic and condensed manner the trends or approaches regularly cited in works in the ETS, with a view to identifying some invariant principles of this research perspective and part of the epistemological questions it raises.

While the question of professionalization can be studied from a multidisciplinary perspective and placing ourselves at different scales of analysis, the concepts and methods presented in this chapter support a way of studying professionalization processes in terms of a "reflection-action" logic (Wittorski 2005, p. 22), closely related to the attention paid to the flows of the subject's activity. In this view, far from only constituting a local and detailed viewpoint, the production of valid knowledge on very small units of activity enables a broader understanding of human work and the transformation forms which accompany it.

9.6. References

Albero, B. (2010). Une approche sociotechnique des environnements de formation. Rationalités, modèles et principes d'action. *Education et didactique*, 4(1), 7–24.

Albero, B. and Brassac, C. (2013). Une approche praxéologique de la connaissance dans le domaine de la formation. Éléments pour un cadre théorique. *Revue française de pédagogie*, 184, 105–120.

Albero, B. and Guérin J. (2014). L'intérêt pour l'"activité" en sciences de l'éducation : vers une épistémologie fédératrice ? *TransFormations*, 11, 11–45.

Albero, B., Guérin, J., Watteau, B. (2019). Comprendre la relation entre influences de l'environnement et activité : questionnements théoriques et enjeux praxéologiques. *Savoirs*, 49, 103–124.

Bakhtine, M. (1975 [1978]). *Esthétique et théorie du roman* (translated by D. Olivier, *Voprosy literatoury i estetiki*). Gallimard, Paris.

Barbier, J.-M. (2000). Rapport établi, sens construit, signification donnée. In *Signification, sens, formation*, Barbier, J.-M. and Galatanu, O. (eds). PUF, Paris.

Barbier, J.-M. (2018). Expérience, transformation de soi, construction du moi, affirmation du je [Online]. Available at: https://www.innovation-pedagogique.fr/article2926.html.

Billet, S. (2001). *Learning in the Workplace. Strategies for Effective Practice*. Allen & Unwin, Crows Nest.

Bronckart, J.-P. and Bulea, E. (2006). La dynamique de l'agir dans la dynamique langagière. In *Sujets, activités, environnements*, Barbier, J-M. and Durand, M. (eds). PUF, Paris.

Brousseau, G. (1998). *Théorie des situations didactiques*. La Pensée Sauvage, Grenoble.

Caens-Martin, S. (1999). Une approche de la structure conceptuelle d'une activité agricole : la taille de la vigne. *Education permanente*, 139, 99–114.

Champy-Remoussenard, P. (2005). Les théories de l'activité entre travail et formation. *Savoirs*, 8, 11–50.

Clot, Y. (1999). *La fonction psychologique du travail*. PUF, Paris.

Clot, Y. (2008). *Travail et pouvoir d'agir*. PUF, Paris.

Clot, Y. (2011). Théorie en clinique de l'activité. In *Interpréter l'agir : un défi théorique*, Maggi, B. (ed.). PUF, Paris.

Clot, Y. and Faïta, D. (2000). Genres et styles en analyse du travail. Concepts et méthodes. *Travailler*, 4, 7–43.

Dejours, C. (1980). *De la psychopathologie à la psychodynamique du travail*. Bayard, Paris.

Dejours, C. (2016). *Situations du travail*. PUF, Paris.

Dewey, J. (1934 [2010]). *L'art comme expérience* (translated by J.-P. Cometti, *Art as Experience*). Gallimard, Paris.

Dewey, J. (1938 [2006]). *Logique : la théorie de l'enquête* (translated by G. Deledalle, *Logic: The Theory of Inquiry*). PUF, Paris.

Durand, M. (2009). Analyse du travail dans une visée de formation : cadres théoriques, méthodes et conceptions. In *Encyclopédie de la formation*, Barbier, J-M., Bourgeois, E., Chapelle, E., Ruano-Borbalan, J-C. (eds). PUF, Paris.

Faingold, N. (2011). Explicitation des pratiques, décryptage du sens. In *Approches pour l'analyse de l'activité*, Le Meur, G. and Hatano, M. (eds). L'Harmattan, Paris.

Felix, M.-C. and Saujat, F. (2015). L'intervention-recherche en milieu de travail enseignant comme moyen de formation. *Raison éducatives*, 19, 201–218.

Filliettaz, L. and Zogmal, M. (eds) (2020). *Mobiliser et développer des compétences interactionnelles en situation de travail éducatif*. Octares, Toulouse.

Filliettaz, L., Bimonte, A., Koleï, G., Nguyen, A., Roux-Mermoud, A., Royer, S., Trébert, D., Tres, C., Zogmal, M. (eds) (2021). Interactions verbales et formation des adultes. *Savoirs*, 56, 11–51.

Galperine, P.-I. (1966). Essai sur la formation par étapes des actions et des concepts. In *Recherches psychologiques en URSS*, Leontiev, A., Luria, A., Spirnov, A. (eds). Les Éditions du Progrès, Moscow.

Habermas, J. (2021). *Pourquoi ne pas lire ?* Première Parallèle, Paris.

Husserl, E. (1905/1964). *Leçons pour une phénoménologie de la conscience intime du temps* (translated by H. Dussort, *Vorlesungen zur Phänomenologie des inneren Zeitbewussts*). PUF, Paris.

Lakatos, I. (1978/1994). *Histoire et méthodologie des sciences. Programmes de recherche et reconstruction rationnelle* (translated by C. Malamoud, J.F. Spitz, and L. Giard, *The Methodology of Scientific Research Programmes*). PUF, Paris.

Laplantine, F. (2003). *De tout petits liens*. Mille et une nuits, Paris.

Lave, J. (1991 [2016]). Situating learning in communities of practice. In *La professionnalisation en formation. Textes fondamentaux*, Wittorski, R. (ed.). PURH, Rouen.

Leontiev, A. (1972). *Le développement du psychisme*. Editions Sociales, Paris.

Leplat, J. (1997). *Regards sur l'activité en situation de travail. Contribution à la psychologie ergonomique*. PUF, Paris.

Linard, M. (1989). *Des machines et des hommes. Apprendre avec les nouvelles technologies*. L'Harmattan, Paris.

Linard, M. (2001). Concevoir des environnements pour apprendre l'activité humaine, cadre organisateur de l'interactivité technique. *Sciences et techniques éducatives*, 8(3–4), 211–238.

Marcel, J.-F., Olry, P., Rothier-Bautzer, E., Sontag, M. (2002). Les pratiques comme objet d'analyse. *Revue française de pédagogie*, 138, 135–170.

Marx, K. (1867 [1969]). *Le Capital. Critique de l'économie politique* (translated by J. Roy, *Das Kapital. Kritik der politischen Ökonomie*). Editions Sociales, Paris.

Maturana, H. and Varela, F. (1980). *Autopoiesis and Cognition: The Realization of the Living*. D. Reidel Publishing Company, Boston.

Mayen, P. (1997). Compétences pragmatiques en situation de communication professionnelle. Une approche de didactique professionnelle. PhD Thesis, Université Paris 5.

Mayen, P. (2013). John Dewey, l'éducation et la reconstruction continue de l'expérience. *Education permanente*, 198, 9–22.

de Montmollin, M. (1961). *Nouvelles perspectives dans l'étude du travail*. Dunod, Paris.

de Montmollin, M. (1974). *L'analyse du travail : préalable à la formation*. Armand Colin, Paris.

de Montmollin, M. (1986). *L'intelligence de la tâche. Éléments d'ergonomie cognitive*. Peter Lang, Bern.

Mouchet, A. (2018). *L'expérience subjective en recherche et en formation*. Presses Universitaires du Septentrion, Villeneuve d'Ascq.

Ochanine, D.A. (1969). Rôle de l'image opérative dans la saisie du contenu informationnel des signaux. *Questions de psychologie*, 4, 209–224.

Oddone, I., Re, A., Briante, G. (1977 [1981]). *Redécouvrir l'expérience ouvrière. Vers une autre psychologie du travail ?* (translated by M.-L. Barsotti, *Esperienza operaia, coscienza di classe et psicologia del lavoro*). Editions Sociales, Paris.

Olry, P. and Bautzer, E. (2003). Diagnostic kinésithérapique. Les conditions d'une coproduction. *Recherche et formation*, 42, 35–50.

Ombredane, A. and Faverge, J-M. (1955). *L'analyse du travail*. PUF, Paris.

Paggetti, M. (2019). Agir et apprendre par le corps : l'activité d'"enquête" du psychomotricien. *Recherche et formation*, 92, 57–69.

Pastré, P. (1992). Essai pour introduire le concept de didactique professionnelle. Doctoral Thesis, Université Paris 5.

Pastré, P. (2011). *La didactique professionnelle, approche anthropologique du développement*. PUF, Paris.

Peirce, C-S. (1935). *Collected Papers*. Harvard University Press, Cambridge.

Piaget, J. (1967). *Biologie et connaissance*. Delachaux et Niestlé, Neuchatel.

Piaget, J. (1974). *La prise de conscience*. PUF, Paris.

Pinsky, L. (1992). *Concevoir pour l'action et la communication : essais d'ergonomie cognitive*. Peter Lang, Neuchatel.

Piot, T. (2006). Les compétences pour enseigner. Contribution à la compréhension de la notion de compétences dans les métiers de l'interaction humaine. HDR, Université de Nantes.

Piot, T. (2017). Faire l'expérience de situations pour (s') apprendre. L'exemple d'une formation d'adultes éloignés de l'emploi comme analyseur d'une pédagogie deweyenne. *Questions Vives*, 27, 77–90.

Rabardel, P. (1995). *Les hommes et les technologies ; approche cognitive des instruments contemporains*. Armand Colin, Paris.

Rabardel, P. (2005). Instrument, activité et développement du pouvoir d'agir. In *Activité, connaissance, organisation*, Lorino, P. and Theulier, R. (eds). La Découverte, Paris.

Savoyant, A. (2006). Tâche, activité et formation des actions de travail. *Education permanente*, 126, 127–136.

Sensevy, G. (2012). *Les sens du savoir. Eléments pour une théorie de l'action conjointe en didactique*. De Boeck, Louvain la Neuve.

Schwartz, Y. (ed.) (1998). *Reconnaissance du travail. Pour une approche ergologique*. PUF, Paris.

Schwartz, Y. (2012). *Expérience et connaissance du travail*. Les Editions Sociales, Paris.

Suchman, L. (1985). *Plans and Situated Action. The Problems of Human/Machine Communication. Intelligent Systems Laboratory, Xerox Palo Alto Research Center*. Cambridge University Press, Cambridge.

Theureau, J. (1992). *Le cours d'action, analyse sémio-logique : essai d'une anthropologie cognitive située*. Peter Lang, Bern.

Thievenaz, J. (2012). L'activité d' "enquête" du médecin du travail. *Recherche et formation*, 70, 61–74.

Thievenaz, J. (2014). L'intérêt de la notion d'"enquête" pour l'analyse du travail en lien avec la formation. *Travail et apprentissages*, 13, 14–33.

Thievenaz, J. (2019a). La théorie de l'enquête de John Dewey. Réexplorations pour la recherche en sciences de l'éducation et de la formation. *Recherche & formation*, 92, 19–38.

Thievenaz, J. (2019b). *Enquêter et apprendre au travail. Approcher l'expérience avec John Dewey*. Raison et Passions, Dijon.

Thievenaz, J. (2021). Conceptions du processus de transmission sur le terrain de la formation des Compagnons du Devoir. *Éducation & didactique*, 15(2), 191–201.

Thievenaz, J., Barbier, J.M., Saussez, F. (eds) (2020). *Comprendre/transformer*. Peter Lang, Bern.

Vergnaud, G. (1990). La théorie des champs conceptuels. *Recherches en didactique des mathématiques*, 10, 133–170.

Vermersch, P. (1994). *L'entretien d'explicitation*. ESF, Paris.

Vermersch, P. (2016). L'entretien d'explicitation une superbe imprudence méthodologique ! Remémoration et explicitation. *Recherches qualitatives*, 20, 559–579.

Vygotski, L-S. (1934 [1997]). *Pensée et langage* (translated by F. Sève, *Myslenie I Rec*). La Dispute, Paris.

Wittorski, R. (2005). Entre travail et formation : la professionnalisation. Habilitation à diriger des recherches. HDR, Sciences de l'éducation. Spécialité formation d'adultes, Paris.

10

Professionalism and the Auto/Biographical Imagination

Linden WEST
Canterbury Christ Church University, UK

10.1. Introduction: auto/biography and professionalism

I suggest, in this chapter, that what we might call the auto/biographical narrative imagination provides a meaningful contribution to the development and understanding of positive dynamics in professional life, and learning and processes of realizing a satisfying professionalism, if never quite in conditions of our own choosing. Professional cultures can be enriched and become more self-confident at a time of profound constraint emanating from neo-liberal and managerialist imperatives and their associated critique or skepticism toward professionalism in the contemporary world. We live at a historical moment when the values and self-knowledge required to respond satisfactorily to clients in good enough ways – what in French is termed *prudence* (see Chapter 1) – is profoundly problematized. I take *prudence* to encompass judgement grounded in an understanding and application of specialist/scientific knowledge, alongside ethical sensibilities, phronesis or practical wisdom and self-knowledge. Fundamentally, it is the capacity to relate to individual clients, in diverse settings, in their singularity, contextual complexity and humanity.

What is meant by an auto/biographical imagination is the capacity to imagine ourselves, empathically, in someone else's shoes and to be able to imagine the world and experience through their eyes, stories and biographies; and to be present, attentive and wise in relation to them. Auto/biography involves the capability to

Encyclopedia of Professionalization,
coordinated by Didier DEMAZIÈRE and Richard WITTORSKI.

draw, reflexively and symbolically, on our own biographies to help illuminate the lives of others. Storytelling lies at the heart of auto/biography: the ability to create spaces which encourage stories and their development, and cultivate the capacity of respectful listening, which includes learning to manage anxieties in self as well as the other. Storytelling helps to create, over time, a good enough relationship with the other, epistemological complexity and interpretative sophistication. It requires, prudentially, a generosity of spirit and time, even in "10 min", heavily constrained cultures of "care". It constitutes a quest, no less, to understand why and how people – professionals as well as their clients – act, suffer and potentially learn, holistically.

Making sense of the complex material we call auto/biography necessitates, in short, symbolic eclecticism and even a kind of relational playfulness. It is not so much about either/or in professional judgement – between science and art, for instance, or technical competence and an auto/biographical imagination (Formenti and West 2018). Rather it is more the spirit of "and also" in ways of knowing and responding. Eclecticism can include "psychosocial" sensibilities in professional life, as in my own work – combining psychoanalytic perspectives with those of a more socio-cultural orientation. It might include systemic sensibilities or, for some, the spirit of pragmatism in a quest for what is good and beautiful in professional life. Knowing, and knowledge, in this pragmatic perspective, is caught, inevitably, in a language game. Language is not synonymous, however sophisticated, with reality; it is a form of representation, which requires reflexive awareness and a capacity to play with diverse perspectives.

A note will be made of how the focus of auto/biographical practitioners and researchers may be on narrative theory, and the rhythms of time and change processes in storytelling and interpretation. Here, an artistic imagination is in play, where professionals working with vulnerable families are themselves enabled, playfully, creatively and prudentially, to bring aspects of their own experience to life, artistically, as a prerequisite for doing so with their clients. The eclectic capacity to play, experiment and to know, in varying ways, lies, I suggest, at the core of a meaningful reinvigoration of professionalism. As does reembodying the idea of professional life as vocation, a calling, rooted in part in our own biographies, a quest for meaning and satisfying values and practice. Such ideas, among clients and other "stakeholders", are easily lost in an era of obsessive economic cost–benefit analysis and quantitative accountability. What is termed the biographical turn in professional lives can, at such a moment, be empowering and epistemologically, as well as existentially satisfying for those at the heart of the process (Apitzsch et al. 2004).

However, a focus on self, subjectivity and auto/biography may be seen as irrelevant, mere messy subjectivism and anecdotalism, a kind of contamination of proper, "objective", "scientific", evidence-based professional practice. There can be profound resistance to framing professional life and development in auto/biographical, narrative and psychosocial terms at all. Scientistic perspectives, we might call them, can still dominate older professions, like medicine, and even haunt newer ones. The nature of being a professional in high status professions like medicine remains overly dominated by scientific, even scientistic assumptions of what counts as valid knowledge, while newer professions can be caught in an objectivist gaze (Furlong 2005; Fook et al. 2016). Scientistic, note, not scientific: the former is more of a normalizing epistemological and methodological imperative, in which a rhetoric of objectivity and quantitative evidence are seen as the key to enhanced professionalism, swept along in the dangerous currents of neo-liberalism and managerialism.

Historically, becoming a professional in high status professions like medicine or the law involved obtaining a licensed autonomy: a contract with the state in which freedom to practice was based on completing a course of extensive scientific and specialist technical training, with relatively limited subsequent external oversight (West 2001, 2009; Furlong 2005). That high status world has of course collided with the discourse of managerialism: if you can measure it, you can manage it, according to the McKinsey dictum (Bogdanich and Forsythe 2019). This is also a way of imposing external, often standardized systems of audit and monetary cost disciplines on professionals and their work. The singularity of the professional relationship with a particular client and their experience is easily lost.

Neo-liberalism is deeply implicated in these trends: its assumptions penetrate to the core of the meso- and micro-levels of professional life and their rationalization. New Public Management Theory (NPMT) trumpets the criteria of economic efficiency, judged through the lens of economic cost–benefit analysis and audit regimes employing standardized, numerical measurement. Champy and Mias, in this volume, make a similar point. NPMT, Champy observes, questions to the core the value of *prudent* nuanced professional practice, focused on singular cases, and a concrete and complex epistemology. We have instead a valorization of quantification and reductive prescription from on high.

Contextual judgment – *prudence* – becomes subsidiary, or highly constrained, even in high status professions like medicine, as well as in professions like health and social care. There can be a manualization of what professionals do, in fields like social work and teaching: the manual will guide you to a "correct" judgement. The *prudence* of professionalism becomes devalued in the technicist imperative

(Fook et al. 2016). The manualization/standardization of professional life carries a real prudential price in the quality of what may be available to clients. Creativity, motivation, judgement and practical wisdom (and the profound potential of the auto/biographical narrative imagination) are compromised (Fook et al. 2016; Fraser 2018).

The price paid can be tragic: in social work, for instance, when working with vulnerable children and families. Vulnerable children have died through a kind of defensive addiction to manuals and the absence of creative, contextual, prudential judgement, as chronicled in various British National Enquires into childcare and social work (Munro 2011). There is evidence of disenchantment in professions like social work, education and medicine with many leaving the professions complaining of stress, burn out and de-professionalization (West 2009, 2016; Formenti and West 2018). In medicine, including general practice, excessive workloads and the continuing neglect of the psycho-emotional dimensions of being a doctor do not help, no doubt exacerbated in the time of Covid (West 2001; Bainbridge et al. 2021).

The decline of status or even trust in many professional communities, and the cultivation of public dissatisfaction with their performance by cost-conscious, neo-liberal politicians and decision-makers is pervasive (Fook et al. 2016). However, there is a different story to tell, even in medicine, when auto/biographical and narrative perspectives find purchase. In fact, narrative and scientific perspectives can be seen as complementary and mutually enriching (Greenhalgh and Hurwitz 1998; West 2001; Formenti and West 2018; Mazzoli Smith 2021). The dots on an epidemiological scattergram are but reductive, if necessary, representations of real, complex people living in families and communities. They have stories to tell, including experiences of illness and pandemics (Greenhalgh and Hurwitz 1998; West 2001, 2016). Auto/biographical research among medics illuminates the problems of purely “scientific” knowing, and the continuing neglect of sociocultural and psychological medicine. I have developed in-depth auto/biographical narrative research and workshops among general practitioners (as well as among other professionals), which illuminate the struggles for meaning, satisfaction and epistemological richness in professional lives (West 2001, 2004; Formenti and West 2018).

There is, in fact, on the margins, in both British and continental European contexts, a growing interest in using auto/biographical narrative methods in professional development and research (and these, or something similar, can be found under labels such as biography, narrative, autoethnography, each with distinct aspects but sharing much in common). This family of methods has gained greater attention across the social sciences (Dominicé 2000; Apitzsch et al. 2004;

Goodson et al. 2017; Formenti and West 2018), even if this has never been mainstream in an "evidence-based" world (Merrill and West 2009; Formenti and West 2018).

10.2. The central role of the European Society for Research on the Education of Adults (ESREA)

I want to chronicle the power of auto/biographical, narrative and interdisciplinary imagination to illuminate the singularities, complexities and diverse ways of knowing required in various professional fields, and of the struggles of many professionals for a more reflexive, aesthetically satisfying, meaningful professionalism; and research, under the auspices of bodies like the European Society for Research on the Education of Adults (ESREA) Life History and Biography Network. Its publications and preoccupations constitute a rich terrain (Merrill and West 2009; Formenti and West 2018; Bainbridge et al. 2021). This is a community engaged in a quest to renew professional life through the cultivation of self/other reflexive, psychological, cultural and epistemological awareness. The network has evoked a shared, ecological vitality, motivating and sustaining professional learners in difficult contexts (Bainbridge et al. 2021). The "whole" of the Network is more than the sum of its parts: its research and professional development roles have influenced practice across Europe, in North America, South Africa, South America and Australasia (Bainbridge et al. 2021).

The scope of this collective imagination encompasses the macro-level of professional life – historical, cultural and discursive forces, like neo-liberalism, that shape cultures and the contexts in which people work. The meso-level of relationships, professional groupings, and cultural preferences. And the micro-level of the most intimate of relationships, and of the workings of inter- and intra-subjective life. Professional perspectives can be shaped by unconscious forces that need to be brought into conscious awareness if lives and practice are to be enriched to the benefit of everyone concerned (Bainbridge and West 2012).

Working life is a prime focus for some researchers, providing opportunities for learning or defensiveness around, for instance, gender and gendered practices and the social unconscious (Salling and Oleson 2020). Henning Salling Oleson and Kirsten Weber have focused on medics, with reference to gender, work and learning. They argue for a new kind of professional identity for general practitioners, beyond a technical model of applying bio-medical knowledge to repair bodily malfunction. There is an urgent task of responding to the messier needs of individual patients, as well as of self; and of really listening to patients and their stories. And being able to draw on a doctor's own life history: of being unable to relate to psychological suffering in an encounter with a migrant, and their child struggling to cope (West 2001; Salling and Oleson 2007).

10.3. A case in point

My own auto/biographic research into professional lives draws, inter alia, on the ideas of Winnicott (1971). Winnicott suggests that psychologists often dwell purely on a person's inner life, separate from context, culture and relationships. Those interested in the sociocultural, on the other hand, often neglect internal life. He calls us to imagine what happens when we are listening to a soothing piece of music, or are absorbed in an idea, or listening deeply to a client; when we have been able to tell a story, however traumatic, because we feel another person is really listening and present to us; or when we observe children are joyfully playing with toys, and creating whole imaginary worlds under the loving gaze of a parent. This is not simply to do with inner and outer worlds, but with the creation of a particular quality of space between people; one neither inner nor outer, but in between, both me and not-me. Winnicott asks us to reflect on an adult or child's enjoyment of living, beauty and play; it maybe, on potential playfulness in professional life, of storytelling and being absorbed in conversation, interpretation and mutual understanding. The idea of a good enough transitional space has shaped my research and professional development workshops.

I want to illustrate the above by reference to a case study of the centrality of care/*prudence* in general medical practice. In one auto/biographical study of general medical practitioners, Jungian-inspired notions of professional and personal individuation came to force. The process required good enough transitional research space, developed over time with a number of doctors. To consider what we can call the feminine and masculine in medicine, for instance, in the case of a doctor called Daniel Cohen (a pseudonym), Daniel talked of the hyper-masculinization of the profession, under evidence-based mantras and managerialism, on the one hand, and a more reflexive, feminine stance in general practice, on the other hand. Crucially, Daniel talked of the biographical roots of his professionalism. Being Jewish and the child of refugees fleeing Nazism was an important part of the story:

> [There was an] "unspoken imperative to succeed and not to rebel…I don't think I had anything that could remotely be described as an adolescence. I think I was just sort of completely studious and privately very distressed and outwardly tremendously successful, like hundreds of Jewish teenagers I knew…".

Daniel came to recognize how he acted as a kind of bridge and a guide for many of his patients, as had an old Jewish general practitioner in the East End of London for his own distressed family. There he said an emotional texture in visits to the doctor, which he now identified as creating safety, containment and good enough space for dialogue. He introduced the story of a suffering Somali woman refugee and her young child. His role was to provide a secure base in a technical but also common-sense understanding: a place to feel looked after, dependably, in a professionally loving way. It was a search for mutual meaning and the incarnation of values of care, which were as important as materialism, if not more. He thought that learning to be human was in fact at the heart of learning to work effectively with others:

> [...] "I remember a therapist saying he would be deeply suspicious of any doctor or psychiatrist who wasn't in the job principally to make themselves better. That if you are not on a kind of journey, trying to understand yourself and make yourself whole then you are probably a bit messianic and so I think that is at the heart. And I hope that as a sort of spin off of that, other people get helped as well".

Daniel saw himself as situated between the truth discourse of mainstream medicine and the uncertainties and messiness of whole people living in what could be a harsh, precarious environment. A subversive synthesis was required, he insisted, by taking what was essential from the medical model but locating this within a person and narrative and biographically focused, even spiritual practice (West 2001, 2004; Formenti and West 2018).

10.4. Doing "educational biography"

Pierre Dominicé's work on "learning from our lives", in Geneva, has been important across Europe in the development of auto/biographical and narrative workshops for professionals (Dominicé 2000). Dominicé co-coordinated the ESREA Network over many years. He used the term "educational biography" as a distinct area of practice and theory building. Educational biography was not primarily a tool for better knowledge of adult development or adult psychology, and did not belong to the domain of psychology or sociology. Instead, it offered insight into what it means for an adult to become more of a knowledgeable, reflexive subject in their professional life.

This is done by generating stories in small groups, and working over time, iteratively, to translate oral accounts into written biographies. Theoretical resources include Paul Ricoeur and his work on hermeneutics. Interpretation has a central role and becomes a shared endeavor as each author's interpretation is submitted to a group of peers for their contributions. Educators benefit from the insights into how people see their learning activities. Writing itself is an act of testimony about our own life, identity and professional experience. The whole process in effect is about learning about learning. Professionals come to realize through their narrative biographies that whatever defenses characterize the journey, they can still become more the author of professional lives (Dominicé 2000).

Michel Alhadeff-Jones (Breton et al. 2018) was a student of Dominicé in Geneva and developed life history seminars at Teachers College, Columbia University, New York, on the theme of "Leadership and self-development: A biographical approach". Here, professionals share significant life experiences in relation to power, authority, control or mastery: orally at first, and then using written narratives, which are submitted to the group for shared interpretation. The aim is to provide students with a "safe" space and time to formulate and question the way they give meaning to experience and to consider the assumptions they make about power dynamics in their lives and in encounters with clients.

When Alhadeff-Jones participated in the Genevan seminars, these lasted for a period of eight months. At Columbia, he had to adjust a process from over 30 sessions to 16 weeks. Such "acceleration" provoked "rhythmic dissonance" (Alhadeff-Jones 2017). Such rigid temporal economies do not work well in processes requiring slower rhythms and time for storytelling, hermeneutics and writing.

Breton et al. (2017) and Breton (2021) and colleagues run workshops at the University of Tours and bring other interpretive resources to bear, including the ideas of Gaston Pineau (Pineau 1983/2020) and the French *Association de Vie Histoire en Formation*. Pineau stresses the importance of considering each story's singularity and subjectivity, but contained within the collective, social, as well as biological life. Pineau draws links between the importance of our relationship to the environment, including its abuse, and broader awareness of the overall health of ecosystems of which we are part, and the way this affects each of us in turn.

To progress in auto/biographical narration, students examine some past moments in order to describe them in words. The activity mobilizes different "acts of language" whose functions merit detailed examination. The professionals have already studied different dilemmas relating to the problems of time. A second perspective is then examined: the different ways by which experience gets translated into words. One of the most significant demands is to find suitable words to express different past moments, at first for self and then to communicate to others. The dilemmas of professional experience can then be expressed in more vibrant language, to be developed into a story.

The whole process is a dilemma for professionals and the practitioner facilitating the seminar. For the professionals, the problem can be threefold: the identification of experiences felt to be valid for inclusion, the choice of words to express these and the anticipated consequences of first-person narratives that may generate shame or inhibition and thus the need for discrimination and judgement.

10.5. More Canterbury tales

Space has been created in Canterbury for auto/biographical narrative workshops among other public sector professionals – teacher educators, career counsellors, health and social workers – using methods derived from the Geneva model (see below) (Merrill and West 2009; Reid and West 2011; Bainbridge and West 2012; Formenti and West 2018; Bainbridge et al. 2021). The work encompasses a search for meaning and wisdom in professional and personal lives (Fraser 2018). The methodology also brings the worlds of literature and textual hermeneutics into illuminating change and resistance to it (Chapman Hoult 2012).

Like Dominicé, Bainbridge (2015) uses the term "educational biography" in his work with professionals, including teachers. He talks of biography being developed by collecting narratives of experience across the lifespan, which are then used to build meaningful links between past experience and professional practice. Educational biographies expose the often-hidden complexity of the interaction between individual experience and the reality of a professional lived-life, or work on experiences beyond "face-value", considering the dynamics of self, other and unconscious processes. The narratives embedded within educational biographies represent the past and present that, when used in the development of education professionals, brings into awareness resistant or motivating behavior, either "stuck-ness" or the possibility for transformation. Narratives themselves constitute good enough transitional space where difficult thoughts and feelings are contained, thought about, and ultimately shaped in more purposeful ways.

Auto/biographical perspectives have similarly been used with nurses and midwives, especially from international backgrounds. Here, cultures of "care" shape how practice is perceived and experienced (Howatson-Jones and Thurgate 2014). Howatson-Jones and Thurgate illuminate the rising concerns at how the body is perceived in healthcare, which has become more impersonal, less compassionate and more technocratic. They have sought to reframe nurse education holistically and have devised new auto/biographical approaches in nursing degree programs. They use workshops and narrative interviews to support people in their developmental journeys, including relationships with selves, each other and organizations. In cultivating sensitivity and awareness of auto/biographical narratives, it becomes more possible to be attuned and responsive to those for whom they care. Howatson-Jones and Thurgate insist professionals can thus help restore more of "the soul of nursing" in their practice.

10.6. Narrative competence

Marianne Horsdal, in Denmark, emphasizes the importance of narrative competence in auto/biographical work (Horsdal 2012). Narrative enables us to make better sense of interactions in changing environments. She suggests that we strive toward authenticity and meaning through narrative endeavors. Composing a narrative is equivalent to receiving a gift from our bodies, which exist in time and space. Every single narrative widens the horizon of self and attentive listeners. The analysis of a narrative encompasses far more than themes. We must examine the order of what is told, its configuration, different voices, the duration of narrated episodes, metaphors used and of interactions, participation and affiliations in various communities of practice. A text is opened to careful, systematic and thorough interpretation. The narrated experience of relationships, in different contexts, illuminates features of formal, non-formal and informal learning, and the problems and possibilities associated with enhancing these in professional (as well as personal) lives.

Other researchers emphasize how narrative work encourages more sophisticated, eclectic understanding among health care professionals (Mazzoli Smith 2021). Mazzoli Smith quotes the writer Charon to draw a distinction between listening for stories and listening to them, the latter being crucial for healthcare practitioners: "... only this kind of narrative listening will hear the connections among body, mind, and self, and disease recognition and treatment cannot proceed, we are beginning to believe, without simultaneous attention to all three" (Charon 2006, p. 67).

The philosophical rationale is pragmatism, not only as paradigmatic positioning, but as a pedagogical resource to unify research and practice. It can be introduced explicitly to orient professional learners to contexts where understanding is difficult or unsettling. Healthcare professionals eventually become more critical of supposed "essentialist" realities, to which they are supposed to owe obedience, moving to a subtler understanding where "There is nothing to be known about an object except what sentences are true of it" (Rorty 2009, p. 55); that the smell of a patient or how they relate to others speaks too of disease. We encounter the revolutionary idea that no-one but us – in what we say and how we choose to engage with the other – is responsible for understanding a patient, from different perspectives. Learners previously unfamiliar with narrative pedagogy, or working with narratives at all, can be liberated by an approach encouraging the idea of more than one truth.

Pragmatism's philosophical anti-foundationalism offers reassurance against the threat of a stark binary between science and more personal forms of understanding. Narrative can reassure healthcare practitioners that different meaning frames help build more satisfying accounts of the "truth" of disease or illness. The discovery of different perspectives on what is beautiful and true helps foster a more vital and diverse ecology of professional learning. The theoretical framing of pragmatism allows, it is suggested, professionals to move more easily beyond dualistic thinking to embrace different understandings of the stories patients tell. Adopting philosophical pragmatism has practical consequences in movement from categorical to eclectic ways of knowing.

10.7. Travelling south

The network has drawn on Southern European perspectives encouraging experiment with auto/biographically orientated artistic workshops. These include the use of art, drawing, theatre as well as storytelling; and the application of systemic theory and different levels of learning – the experiential (from direct experience), the practical or applied knowing; the presentational (a kind of artistic rehearsal process), as well as the more conceptual and abstract propositional domain in John Heron's framing (Del Negro 2021). Gregory Bateson's ecology of mind is important as a theoretical and methodological resource in the systemic work of Laura Formenti in Milan, who has also co-coordinated the network (Formenti 2014; Formenti and West 2018). Here, the framing of workshops encourages connectedness within an ecological system: between the professional and their clients and the quality of relationship between them and a wider world. In families, one member is best understood by reference to patterns within the larger unit: and by extension, a wider ecology of relationships, opportunities, cultural as well as material constraints.

There is a focus on the interplay of perspectives between professionals and their clients, and the influence of family histories, consciously or not. These can include myths surrounding birth in families. Professionals bring all kinds of birth myths into practice, which can be identified in professional workshops, using a literary approach. So, sessions begin with readings from the Ancient Greeks, telling stories of the birth of the Gods. Then, each participant is invited to write stories of their own myth of birth, such as the late, awkward or unexpected child.

Stories are read aloud and developed. Later, there can be collective discussion about family myths: self-fulfilling prophecies, destinies and agency. The concept of *daimon* as cited by Arendt (1959), and the Jungian analytic psychologist James Hillman (1996), is used to signify the mysterious role of an individuality or presence that can be with us from birth, shaping as much as it is shaped (Formenti 2014). Learning auto/biographically what it was to be a child, a parent or grandparent is considered essential in working with families. What is noticed in family dynamics is how professional perspectives are inevitably shaped by their own familial experience (Formenti 2014; Del Negro 2021).

The idea of an artistic imagination is important in Formenti's (2017) work with family educators. She invites participants to connect with lived experience in families and to playfully explore this. Aesthetic representation, or presentational knowing, is one of four kinds of knowledge, which includes embodied experience, intelligence and deliberative action over cycles. In a two-year project with family tutors working with parents whose children have been fostered, Formenti invites them to explore, in depth, their own representations of work and role, and of how families and stories of accidents and other events provoke deep emotions, even haunting qualities in family life, and thus dilemmas that must be addressed.

A family tutor is an adult educator who seeks to create space for different voices, and for learning and agency within families. Their mandate – given by the state – is to intervene with parents defined as socially, psychologically and/or physically vulnerable, and who need to be "trained" and/or controlled. So, their work with parents can be overly instructive, riddled with deficit assumptions and prone to imposing pre-defined models of parenting skills. Cooperative inquiry becomes a space for new possibility: at each meeting, participants draw, make poetry and do creative writing, or use photographs and children's books, to let the imagination take the lead. Some meaningful patterns can begin to emerge from conversation and inspire deliberate action toward more effective and ecological ways of being and working. Workshops evoke an epiphany when professionals feel the *desire* to practice similar approaches among the families with whom they work.

Art can "set stories in motion" and help re-connect subjects' own experience with educational work. The group of family tutors used the "gaze" metaphor to consider new epistemological positionings: "enamored, complex, acute, and contagious" (Formenti 2017). The metaphor of the gaze helps avoid the risk of trivializing the other and stigmatizing families. Step by step, they build a shared repertoire of rich image, words, actions and representations. There have been public conferences at the end of projects where family tutors decide not only to describe their work, but to engage the public by using the voices of parents and children to generate and explore the role of personal metaphors in lives; metaphors such as "finding a shared rhythm", "building bridges", "cultivating the garden with patience and desire" and "searching for beauty". These show photographs of dance, architecture, lush nature and art. For them, "being professional" involves communicating some of the depth and beauty, the difficulties and challenges, of professional life to wider audiences. The families are there, alongside various stakeholders, welfare officers, decision makers and professionals. All can be emotionally touched and provoked by the representation. The possibility for a deepening dialogue and mutual understanding is created.

Poetic qualities of communication and authentic forms of participation create new ways of telling stories and understanding selves: and the motivation to take more responsibility and create more and better emotional space in professional practice. Tutoring, as in any other educational activity, is circular, while interdependence is to be celebrated. Hope becomes embodied not as mere optimism or positive thinking, but in the auto/biographical motivation to engage in the complexities of family life, develop new professional skills and to internalize good theory and new methods. Conflict and error are unavoidable, but Formenti's idea is not so much to add "new tools to the educator's kit", but to overcome the functionalist metaphor of a toolbox separated from the bodies, lives, minds and souls of educators and their clients.

10.8. Imagining the future

There are clusters of young researchers exercising the auto/biographical imagination in new aesthetically satisfying and interdisciplinary ways, bringing renewed promise for the future. A common finding among them (Rothe 2019; Del Negro 2021) is that professionalism encompasses deeply personal dimensions in a complex conscious and unconscious, cognitive and emotional relational dynamic. Professionals can feel constantly subjected to external standards and disempowered by audit but must negotiate who they are and what they do within these parameters. Their actual encounters and the uncertainty surrounding them challenges overly technocratic and instrumentalist discourse, while indicating the importance of eclectic ways of knowing in professional practice. Researcher/practitioners like Rothe and Del Negro argue for a holistic re-conceptualization of professionalism as necessarily both personal and professional. They use the idea of "reflexive irritation" as an element in professional life and as a basis for professional development.

These researchers generate diverse insights into the dilemmas of professional practice: how, for instance, we may be storied by the power of objectivized, overly abstract "academic" or traditionally scientific ways of knowing, and require good enough space for other epistemological frames to develop. By using stories, drawing and poetry to illuminate reflexive irritation and related dilemmas in professional life, Antje Rothe takes us to the core of dilemmas for early childhood educators. One educator feels space is colonized by a manager's demands to intervene more, when what is really required is greater freedom and space for children and parents (and professionals) to feel and think together. This mirrors the professional's own biography as a child and the lack of space to be, feel and think, in relation to her parents.

The same early years' educator tells Rothe how she grew up in a home characterized as very protective, expecting her always to be proper and well-behaved. Her father was a policeman and her mother a housewife. She remembers that she identified with her father and his profession. She chose him as a role model and at first aimed to become a policewoman. The relationship with her mother she remembers as being more confrontational. She recalls that both her parents tried to protect her from harmful experiences, which she sees as a consequence of her father's profession.

When she began her training, she moved into a shared apartment with her friends. She calls this her "escape", pointing out that this was a conscious breaking away from the overprotective home. This phase is highly significant for her because it took courage to distance herself from her parents and the opportunity provided to know herself in new social situations. In this phase of "creative metamorphoses of biographical identity" (Schütze 2007), she experienced a shift in mind-set. Various insights and convictions emanate from this aspect of her personal history. She talks passionately about gender sensitivity and how she aims to overcome the constraints of traditional gender roles in everyday experience. She talks of how she encourages boys when they reveal curiosity about wearing a dress, while at the same time encouraging girls not to hide their feelings but to express them openly and loudly. Childhood experience of being a "well-behaved" girl, and conflicts with her parents, are auto/biographically significant in this regard.

Del Negro's participatory design workshops with professionals have included six monthly narrative sessions with small co-operative groups in the University of Milano-Bicocca (Italy) and at Canterbury Christ Church University (UK). These were run simultaneously from January to June 2015, with a follow up after three months. Nine participants in Milan and six in Canterbury engaged in metaphoric and embodied cognition, following Heron's (1996) steps of knowledge – the sensuous ("experience"), imagination/intuition ("presentation") and rational understanding ("proposition") – are woven together to create novel thinking and lines of action.

In one meeting, there was a reading from Borges's (1964) "The circular ruins" – a short "pedagogical" story about a magician giving life to a man. It prompted a clash of epistemologies, no less, between someone called Dilbert (highly rational, who stressed the need for professional accountability) and Vanessa, an art-therapist, who sought to avoid, or downplayed the importance of this.

> Dilbert: To have that level of illusion is paralyzing. How can we operate within that then?
>
> Vanessa: Oh now, I think it frees us.
>
> Dilbert: Well free to what? What can you establish, what can you then substantiate? How can you operate when you have nothing to relate to?

Dilbert eventually drew a picture of a mentoring figure whose hands burn holding books, and he wrote auto/biographically about two halves of his life as a lecturer and boxer. Two months after the end of the research workshops, Del Negro invited the participants for a follow-up session in Canterbury and Milan. She proposed a creative activity that consisted of making an object with paper and colors, which could represent what each person learnt from participating in the research. Dilbert attempted to reflect on his values within the functionalist discourse he felt professionally subjected to. The drawing of his childhood household entitled "The unwritten narrative" links the personal to the professional. Dilbert's constrained learner identity as a child, facing a difficult situation at home, links with a present identity as a teacher and his struggle to imagine more eclectically – beyond standard, prescribed curricula – both for himself and students.

Other young researchers (Kirchgaesser 2020) combine auto/biography with social sculpture, film making, drawing, music to explore transitional processes in personal as well as professional lives, in pursuit of meaning, integration and even soul work. Their practice-based activities in "social sculpture" – a phrase coined by German artist Joseph Beuys (1921–1986) to designate an expanded concept of art. Emancipatory and transformative in nature, it involves developing the perceptive, imaginative, reflective and communicative capacities associated with art and applying these to navigating life itself. This implies living with the attitude of an affected, socially engaged researcher and professional: one who, with curiosity and care, seeks to understand and where possible improve the world in which they find themselves, confronting both internalized and external forms of oppression. The interest in social sculpture intersects with questions around the scope of human agency, as well as what helps and hinders creativity, learning and a developing humanity. An eclectic theoretical perspective informs the work: transformative learning theory, psychoanalysis, pragmatism, feminism, queer and systems theory as well as social sculpture.

10.9. Conclusion: a wider world, constraints and new opportunities

Professional lives have become more complex in a world where notions of professionalism are deeply contested. The imperatives of care and *prudence* in professional life are compromised by regimes of regulation and accountability as well as evidence-based practice. Time for prudential care is of the essence as narrowly economistic measures of effective practice are dominant. There are also challenges to auto/biographical and narrative perspectives in research, as well as professional life, which may be especially strong in the Anglophone world.

However, countervailing forces are also at work. The auto/biographical, narrative imagination has created space in diverse professional contexts – in nursing, health care, medicine, teaching, early years education, social care and social work, for instance. So as to generate a growing body of research into the importance of *prudence* and phronesis in professional life: creating transitional space for stories and for greater attention to be paid to nuance and singularity, whatever the time and resource constraints, as a basis for more informed and satisfying judgement. The price paid for an absence of such *prudence* is found in bad clinical and contextual judgement; and even creative paralysis – born of the fear of taking risks – that can lead, in extremis, to the death of children. These matters are, in such terms, about life and death. A body of context-rich, humanly sensitive practice illuminates the inadequacies of much conventional professional training or of top-down notions of "formation" and managerialism, and one-dimensional interpretations of evidence. Eclecticism and *prudence* scream for recognition and space.

Professionals need to know, imagine and practice holistically, even in a harsh neo-liberal world where they struggle with fewer resources and pressures of time, and where only particular forms of reductionist knowledge count. The failure to recognize professionals as experiencing, creative subjects with life experience to draw on, and judgement to exercise, can be disabling. Not least within professional contexts marked by rapid change, technological development, demands for more from less. More, not less, space is required for auto/biographies, storytelling, cultural and personal reflexivity, and diverse ways of knowing. Space to help re-humanize and re-invigorate professional life and the quality of encounters with clients. This is an urgent, essential, ethical and ontological necessity.

Ironically, pressures of accountability are creating new opportunities to enhance the quality of professional life. If many pressures are top-down, some are bottom up. They include the demand for diverse stakeholders to be given voice, which can include clients from the cultural and political margins, like the Somali woman refugee and her child. Such voices might help refocus attention on the prudential qualities at the heart of good professionalism: that make a role a vocation, not just a job; vocation as a calling in which we feel compelled to engage with our own biographies as part of attending more satisfactorily to those of clients. To appreciate, for example, the role of a doctor in our own family's history and the importance of offering a bridge and being fully present, here and now as well as there and then. Or perhaps when working with troubled families in marginalized communities, remembering some of our own difficulties of talking about trauma and distress. Or of learning, once more, to play artistically and imaginatively to challenge myths that haunt families and children, and how new ones can be created in life-enhancing ways. Maybe to relearn the idea that finding satisfaction, once more, in a vocation, is essential to managing stress and inevitable disenchantment, and even the temptations to quit when management feels distantly oppressive and what we are instructed to do makes little or no sense.

10.10. References

Alhadeff-Jones, M. (2017). *Time and the Rhythms of Emancipatory Education. Rethinking the Temporal Complexity of Self and Society*. Routledge, London.

Alhadeff-Jones, M., Breton, H., West, L. (2018). The practice and dilemmas of life history seminars; working at the edge to foster individual and collective transformation. In *International Transformative Learning Conference*, New York.

Apitzsch, U., Bornat, J., Chamberlayne, P. (eds) (2004). *Biographical Methods and Professional Practice. An International Perspective*. The Policy Press, Bristol.

Arendt, H. (1959). *The Human Condition*. Doubleday, New York.

Bainbridge, A. (2015). *On Becoming an Education Professional: A Psychosocial Exploration of Developing an Education Professional Practice*. Palgrave, London.

Bainbridge, A. and West, L. (2012). *Psychoanalysis and Education: Minding the Gap*. Karnac, London.

Bainbridge, A., Formenti, L., West, L. (eds) (2021). *Discourses, Dialogue and Diversity in Biographical Research. An Ecology of Life and Learning*. Brill/Sense, Leiden.

Belenky, M.F., Clinchy, B.M., Goldberger, N.R., Tarule, J.M. (1986). *Women's Ways of Knowing: The Development of Self, Voice and Mind*. BasicBooks, New York.

Bogdanich, B. and Forsythe, W. (2019). How we have reported on the secrets and power of McKinsey and company. *New York Times. Times Insider*, February 26.

Borges, J.L. (1964). The Circular Ruins. In *Labyrinths: Selected Stories and Other Writings*, Yates, D.A. and Irby, J.E. (eds). Penguin Books, London.

Breton, H. (2017). Expériences du temps et récit de soi au cours de l'activité biographique en formation d'adultes. *Chemins de formation. Décélérer pour apprendre ?* 21, 45–57.

Breton, H. (2021). Narrative regimes. An alliance between descriptive phenomenology and biography. In *Discourses, Dialogue and Diversity in Biographical Research. An Ecology of Life and Learning*, Bainbridge, A., Formenti, L., West, L. (eds). Brill/Sense, Leiden.

Chapman Hoult, E. (2012). *Adult Learning and la Recherche Feminine*. Palgrave Macmillan, London.

Charlot, B. (1997). *Du rapport au savoir : éléments pour une théorie*. Anthropos, Paris.

Charon, F. (2006). *Narrative Medicine: Honouring the Stories of Illness*. Oxford University Press, Oxford.

Del Negro, G. (2021). Profession reimagined: Tackling adult educators' alienation through multimodal ways of knowing. In *Discourses, Dialogue and Diversity in Biographical Research. An Ecology of Life and Learning*, Bainbridge, A., Formenti, L., West, L. (eds). Brill/Sense, Leiden.

Dominicé, P. (2000). *Learning from Our Lives*. Jossey-Bass, San Francisco.

Fook, J., Collington, V., Ross F., Ruch, G., West, L. (eds) (2016). *Researching Critical Reflection: Multidisciplinary Perspectives*. Routledge, London.

Formenti, L. (2014). The myth of birth: Autobiography and family memory. In *Embodied Narratives, Connecting Stories, Bodies and Ecologies*, Formenti, L., West, L., Horsdal, M. (eds). University of Southern Denmark Press, Odense.

Formenti, L. (2017). *Formazione e transformazione. Un modello complesso*. Raffello Cortina, Milan.

Formenti, L. and West, L. (2018). *Transforming Perspectives in Lifelong Learning and Adult Education: A Dialogue*. Palgrave Macmillan, London.

Fraser, W. (2018). *Seeking Wisdom in Adult Teaching and Learning: An Autoethnographic Inquiry*. Palgrave Macmillan, London.

Furlong, J. (2005). New labour and teacher education: The end of an era. *Oxford Review of Education*, 31, 119–134.

Goodson, I., Antikainen, A., Sikes, P., Andrews, M. (eds) (2017). *The Routledge International Handbook on Narrative and Life History*. Routledge, New York.

Greenhalgh, T. and Hurwitz, B. (eds) (1998). *Narrative Based Medicine, Dialogue and Discourse in Clinical Practice*. BMJ, London.

Heron, J. (1996). *Co-operative Inquiry: Research into the Human Condition.* Sage Publications, London.

Hillman, J. (1996). *The Souls' Code in Search of Character and Calling.* Warner Books, New York.

Horsdal, M. (2012). *Telling Lives. Exploring Dimensions of Narratives.* Routledge, London.

Howatson-Jones, L. and Thurgate, C. (2014). Biographical learning: A process for recovering the soul of nursing. In *Embodied Narratives, Connecting Stories, Bodies and Ecologies*, Formenti, L., West, L., Horsdal, M. (eds). University of Southern Denmark Press, Odense.

Kirchgaesser, I. (2020). For the love of reality; social sculpture as self-experiment. PhD Thesis. Oxford Brookes University, Oxford.

Laing, R.D. (1967). *The Politics of Experience and the Bird of Paradise.* Penguin Books, London.

Mazzoli Smith, L. (2021). Beyond truth. A pragmatist approach to narrative pedagogy in professional learning for healthcare professionals. In *Discourses, Dialogue and Diversity in Biographical Research. An Ecology of Life and Learning*, Bainbridge, A., Formenti, L., West, L. (eds). Brill/Sense, Leiden.

Merrill, B. and West, L. (2009). *Using Biographical Methods in Social Research.* Sage, London.

Munro, E. (2011). Munro review of child protection: A child-centred system. Report, Department of Education, London.

Pineau, G. (1983/2020). *Produire de sa vie : autoformation et autobiographie.* Téradère, Paris.

Pineau, G. (1986). Histoire de vie et reconnaissance des acquis : éléments d'une méthodologie collective et progressive. *Education Permanente*, 83–84, 139–146.

Reid, H. and West, L.R. (2011). "Telling tales": Using narrative in career guidance. *Journal of Vocational Behavior*, 78(2), 174–183.

Rorty, F. (2009). *Philosophy and the Mirror of Nature.* University Press, Minnesota.

Rothe, A. (2019). *Professionalität und Biografie: Eine qualitative Studie zur Bedeutung biografischer Erfahrungen für die professionelle Identität frühpädagogischer Fachkräfte.* Beltz Juventa, Weinheim.

Salling Oleson, H. (2007). Professional identities: Becoming a general practitioner. In *Using Biographical and Life History Approaches in the Study of Lifelong Learning: European Perspectives*, West, L., Alheit, P., Siig Andersen, A., Merrill, B. (eds). Peter Lang, Frankfurt.

Salling Oleson, H. (2020). *The Societal Unconscious: Psychosocial Perspectives on Adult Learning.* Brill/Sense, Leiden.

Schutze, F. (2008). Biography analysis on the Empirical Base of the Autobiographical Narratives: How to Analyse Autobiographical Narrative Interviews. Part I. *INVITE - Biographical Counselling in Rehabilitative Vocational Training*. Further Educational Curriculum. EU Leonardo da Vinci Programme [Online]. Available at: http://www.uni- magdeburg.de/zsm/projekt/biographical/1/B2.1.pdf.

West, L. (2001). *Doctors on the Edge: General Practitioners, Health and Learning in the Inner-City*. Free Association Books, London.

West, L. (2004). Doctors on an edge: A cultural psychology of learning and health. In *Biographical Methods and Professional Practice: An International Perspective*, Chamberlayne, P., Bornat, J., Apitzsch, A. (eds). Policy Press, Bristol.

West, L. (2009). Really reflexive practice: Auto/biographical research and struggles for a critical reflexivity. In *Beyond Reflective Practice. New Approaches to Professional Lifelong Learning*, Bradbury, H., Frost, H., Kilminster, S., Zukas, M. (eds). Routledge, London.

West, L. (2016). Critical reflection? Auto/biographical narrative enquiry and illuminating professional struggles in distressed communities. In *Researching Critical Reflection: Multidisciplinary Perspectives*, Fook, J., Collington, V., Ross, F., Ruch, G., West, L. (eds). Routledge, London.

Winnicott, D.W. (1971). *Playing and Reality*. Routledge, London.

Conclusion

Manufacturing of Professions, Production of Professionalities and Rise of Professionalism: Interdependence and Reciprocal Transformations of Individuals, Collectives, Organizations and Environments

Richard WITTORSKI
CIRNEF Laboratory, University of Rouen, France

We have reached the end of this encyclopedic book devoted to the study of the field of professionalization, or rather professionalizations. The main intention of the following text is to put forward the hypothesis that far from being separate and functioning autonomously, the three significations of professionalization which structure the encyclopedia are strongly articulated. They reveal particular configurations in the sense of Elias (1987), meaning, close interdependence systems. Here, we will develop the idea that professionalization refers to "space-time" or singular configurations (forms) articulating (putting into debate-tension) varied combinations of projects from professional groups, organizations and individuals in (socio-politico-economic) environments, which have their own characteristics.

As we mentioned in the introduction to this work:

> Work challenges, broader political and social challenges, challenges raised by professional groups and training challenges are closely linked. To talk about professionalization in the field of training, for example, necessarily involves evoking at the same time and in an inseparable way the transformations of work and the broader environments, seeking to understand the relationships of reciprocal transformations between individuals, collectives, organizations and environments.

The following text will be organized into two parts. The first part will offer a synthesis of the three main significations of professionalization as exposed in the three parts of the encyclopedia, highlighting the social and disciplinary logics specific to each: the challenges of the professions with the dominant light of the sociology of professions; the challenges of work in the dominant light of the sociology of work and organizational sciences; the challenges of training with the dominant perspective of education and training sciences. The second part will, in some way, offer an opposite view to the first one by emphasizing interdependence links which in reality unite these three logics of professionalization. It will develop the idea according to which we are rather witnessing an "overall movement" reflecting a new form of social and organizational governance leading to another conception of the "professional", of training… This is accompanied by reciprocal transformations acting on the "major actors" involved (environment (socio-politico-economic), the state, professional groups, organizations, collectives and individuals). The conclusion will close the discussion by identifying some blind points (or little-studied to date) and favoring a deeper understanding of the field of professionalization.

C.1. The study of the three professionalizations: different disciplinary spaces; meanings, challenges and modalities that stand out

The encyclopedia is structured around three main meanings of the term professionalization. We can say that it is difficult to speak of "professionalization in general", but we should rather talk about "professionalizations": the manufacturing of professions, the production of professionalities and the rise of professionalisms do not apparently arise from the same logics of professionalization actors nor from the same dynamics, at first sight.

Three meanings	Dominant challenges	Main "promoters"	Terms frequently observed	Disciplinary fields
Professionalization profession	– Establishing a new sector of activity – Affirming the existence of a profession and having it recognized to improve the statuses and means of existence	– Actors sharing the same activities – Actors campaigning for the development of a new field	– Setting up collective instances (e.g. professional associations and orders) – Institutional negotiations – Rhetoric on activities – Constitution of benchmarks and certifications within the framework of negotiations with the state, with a view to having the profession recognized	– Sociology of professions and professional groups
Professionalization professionalism	– Improving work efficiency For example, through the promotion of effectiveness criteria for thinking about the work of others – Operating and managing change	– Organizations (administrations, companies) which foster a change project or improvement about work – Work collectives expressing professionalism standards more or less formally	– Issuing new work standards – Defining differently what a "professional" is (e.g. promoting skills relating to versatility) – Transforming the terms of work and integrating training into work (on-the-job training)	– Sciences of work and organizations (sociology of work, social psychology and work, management sciences)
Professionalization professionalities	– Preparing newcomers to a profession in the image of said profession	– Education and training environments – Professional environments concerned – apprenticeship programs	– Place of skills and benchmarks	– Training sciences

Table C.1. *Different disciplinary spaces, meanings, challenges and modalities*

When we study the social uses of the term professionalization, and therefore the discourses and social practices, we see that it is not used for the same reasons nor by the same actors (companies, professional groups, individuals, state, etc.), thereby translating different "meanings". For each of them, we can identify a unique "system of links" between the challenges to be surmounted, the actors-promoters and the methods implemented by professionalization, as shown in Table C.1. This can be read row by row, identifying for each signification of professionalization, the dominant challenges corresponding to it, the main "promoters" (in the sense of the actors who are at the origin of the term's use), the modalities frequently observed and the disciplinary field which particularly studies it, even if there are numerous points of meeting between the works (marked by discipline), which shed light on each of the meanings, for example, between the sociology of work and training sciences.

Let us take a look at each of these three meanings, offering an example each time to illustrate the dominant logics at play.

C.1.1. *"Professionalization-profession" studied by the sociology of professions and professional groups*

"Professionalization-profession" in the sense of the affirmation of professions and professional specificities has largely been worked on by the sociology of professions and the sociology of professional groups. The social dynamics of the organization of professionals to have their activity recognized have been the subject of a strong conceptualization, sometimes also having the status of "social proposal" (the ideal of the liberal profession driven by a single desire for altruism and for functionalists). As the texts in the first part of the encyclopedia show, there are numerous debates between these approaches, particularly between those which advocate the ideal of a completed profession and the more recent works which highlight heterogeneous and continuous processes, particularly in Europe.

Here, we see close links between the challenges of professionalization, the promoters of the dynamics observed and the methods implemented. Thus, the dominant challenge is the better recognition of the activity to improve its status and practice conditions. Promoters are often professionals or actors who share the same activity and have every interest in seeing it better recognized. The modalities implemented often involve a rhetorical formalization regarding the professional activity by highlighting its specificities, as well as a negotiation activity with legal recognition bodies via the creation, for example, of professional associations. This also concerns emerging sectors of activity, as will be shown in the following example.

Let us take the example of the organization of the direct home employment sector in France, which shows a "professionalization-profession" dynamic. Direct employment at home concerns the use of the universal service employment check (*Chèque emploi service universel* (CESU)) to pay an employee working at home for a few hours for childcare, cleaning, gardening activities, etc. We then speak of direct employment because the individual pays the employee directly, without going through an intermediary organization. This field of activity is developing strongly.

It has gradually become structured in France thanks to initiatives initially taken by a handful of direct employers:

– On the one hand, the creation of the Federation of Individual Employers (*Fédération des Particuliers Employeurs* (FEPEM)) defining, together with employee representative organizations, the rules which govern the relationships between employers and employees at home. FEPEM also developed a regular communication and exchange activity with the French Ministry of Labor to recognize the specificities of the sector and to propose regulations.

– On the other hand, the creation of a training institute for employees involved in direct employment at home (IPERIA), gradually working toward the structuring of employee activities deployed at home into several métiers (life assistant, family employee, etc.), the establishment of certifications and training associated with these métiers and their recognition by professional training authorities.

– Finally, the recent creation (2020) of a home foundation and a home university, so as to finance research work entrusted to university laboratories. The purpose of research is to better understand the activities and professional skills implemented by employees at home, with a view to consolidating the métiers associated with such expertise, identified by research.

This example shows that "professionalization-profession" involves several combined modalities, in particular the creation of collective bodies with a triple activity for formalizing the existing and developing a rhetoric about the sector; negotiation/recognition with legal bodies (the state, specifically) enabling the attribution of statuses, training engineering and training in the métiers identified.

C.1.2. *"Professionalization-professionalism" studied by work and organizational sciences*

"Professionalization-professionalism" in the sense of the rise of new demands at work was the subject of the second part of the encyclopedia, and was mainly tackled by the sociology of work, work psychology and management sciences. The knowledge produced in these disciplines about professionalization tends to highlight the appearance of new forms of governance at work, often based on the promoted model of an "autonomous and responsible" professional accompanying this rise in professionalism. One of the questions with high social stakes (because it is at the heart of the definition of professionalism), concerns the discussion (who and how?) about the definition or redefinition of "a job well done". Such discussion is sometimes unbalanced between professionals and institutions. These debates often reflect tensions between (reform) institutional projects, collective projects (maintaining the contours of professionals) and individual projects.

Here, we see again the close links between the challenges of professionalization, the promoters of the dynamics observed and the methods implemented. However, these links are different from those observed in a "professionalization-profession" logic. The dominant challenge no longer involves obtaining recognition for a specific activity, but better optimizing work efficiency or managing a major change. The observed promoter is no longer a professional group, but often an organization (company, association, public prosecutor) which drives this project for efficiency and/or change, and enacts new standards at work, defines in a different way what a "professional" is (promoting skills related to polyvalence, for example), transforms work contexts and uses approaches to developing skills on-the-job, in a work situation. This injunction to professionalism brings to the forefront the central question of work autonomy, the challenge of struggles within organizations between the managerial project and the defense of professional autonomy (autonomy *at* work, but also *regarding* work) through professional collectives summoned in the defense of the métier's historical standards, or the recognition of professionalism in action, built in the individual and collective exercise of work. This recognition can lead to steps to formalize this *bottom-up* professionalism with a view to making organizational standards reflect a form of compromise between the managerial approach and the workers' approach to efficient and quality work.

Let us take the example of a Canadian textile company located in Estrie (eastern canton of Quebec), with which we carried out research work for three years at the beginning of the 1990s (study by observations and interviews of a scheme for analyzing the work carried out by the workers themselves in a context of organizational change). This company was facing challenging economic difficulties in its sector. It relied on a Taylorian-type organization of production and had a long tradition of shift work. At the beginning of the 1990s, the factory was plagued by profitability difficulties: the quality of production was not sufficient and market shares were gradually decreasing. The company quickly decided on a sale to

American and Canadian shareholders. With the arrival of new owners, a policy of rapid change was put in place with a view to reducing production costs and regaining profitability. The company therefore went through a process which can be described as an attempt to emerge from a situation of economic crisis by betting on remobilizing all categories of employees around an ambitious change project. It is the Taylorian organization model that management wished to destabilize by replacing it with a more flexible organization model. The company committed itself to a total quality program, which aimed to transform usual working practices.

It also modified work organization, changed several machines and reviewed the usual remuneration arrangements. The total quality program implemented had the central goal of making each operator personally responsible for the quality of their work. The implementation of this program included a significant training component for several categories of factory employees. On the one hand, it was necessary to disseminate the key principles of quality management through several training sessions. Then, the foremen all participated in training aimed at enhancing their presence at the workshops: replacing the traditional role of "order giver" or "commission agent" by that of team leader. The accompanying measures for this total quality program modified the organization chart and the status of certain staff: the hierarchy at each workshop included a workshop foreman and team foremen.

The company's management defined three criteria which became new goals to be achieved by workers: improving the quality of the work produced, safety and productivity. These new work standards were accompanied by a vast work analysis system by the workers themselves, with a view to formalizing the new ways of working in booklets and optimizing the three defined criteria.

This example illustrates the "professionalization-professionalism" logic, whose challenge was to attain the fulfillment deemed necessary for rapid change. The initiator of the approach was the company, which set up new benchmarks regarding what, in its terms, constitutes "an efficient and a well done job".

C.1.3. *"Professionalization-professionality" studied by the training sciences*

"Professionalization-professionality" in the sense of creating a professional future has been specifically studied by education and training sciences.

The third part of the encyclopedia indicates that the knowledge produced about professionalization emphasizes several dimensions:

– the frequent use of (work-training) apprenticeship programs that raises the question of the design of training–work relationships (applicative versus iterative design?);

– the place of learning in a work situation (how is learning built?);

– the use of an activity analysis type of approach, not only to understand professional learning but also to promote it; the modalities and itineraries for the construction of subjects via the biographical approach.

Once again, we notice the close links between the challenges of professionalization, the promoters of the dynamics observed and the methods implemented, but these links are different from those observed in a "professionalization-profession" and a "professionalization-professionalism" logic. Thus, the dominant challenge is no longer to obtain recognition for an activity or to optimize work efficiency or even manage major change, but to train newcomers in a profession. The promoter is no longer a professional group or a company, but a training institution that often sets up an apprenticeship program backed by a skills benchmark reflecting a certain image of the métier. Diploma courses (generally long) may be involved, based on métier diplomas negotiated between the state, representatives of professional groups and social partners. As part of the accompaniment of the unemployed in their return to work, there may also be more or less certifying training programs (rather short), with a view to supporting the recruitment of professionals in activities and métiers showing certain "shortages".

In this third sense, particularly within the frequently used apprenticeship programs, we see a diversity of professionalization paths (Wittorski 2007 and 2016), which often combine work and training experience. In this case, professionalization takes place both during training and work situations, or even in situations allowing for the articulation between experience produced at work and the one produced during training:

– Professionalization through formal training situations: training off-the-job often enables the acquisition of a certain amount of knowledge useful for acting while practicing our métier. For example, it is essential for a lawyer to learn law before practicing their profession, because they cannot discover this during the professional activity, "on-the-job". It is often said that training (outside of a work situation) is one of the preferred paths to professionalization for the acquisition of formal knowledge (in the sense of putting something into words or in writing), which then organizes the practice of the métier. The same is true for a doctor, who must learn how the body works before practicing the profession. One of the dominant schemes here is training off-the-job, for example, at school, college and university. The knowledge acquired in that context constitutes what we can call "formal expertise".

– Professionalization on-the-job: the exercise of work can also lead to the development of professional learning useful for the practice of a métier. As numerous studies have shown, however, this learning, which is sometimes called action or experiential knowledge, often remains "incorporated" in the sense that it is attached to action and situations and not highly transferable (communicable to others or reusable in other situations). Many professionals build their métier in this way, especially self-taught people. One of the dominant forms here is the practice of work, especially if it regularly presents novelty characteristics. It may also involve setting up simulation or role play situations (learning by simulation or role play, in other words, learning by doing off-the-job and in a "security" context: forum theater, etc.), project groups or problem-solving groups (collective professional learning by working with others). The learning produced here constitutes what we can call "expertise in action".

– Professionalization through the combination of learning on-the-job and learning in a training situation: professionalization can also take place through alternating between moments of work (work situation) and the analysis of the experience produced at work (in a training situation), combined with possible additional training contributions according to the needs identified from the analysis of the professional activity implemented. These situations combine work and training and encourage the development of knowledge about and for action (not only do they constitute formalizations of the activity performed, but they also identify new possible avenues for action). They constitute what we could call "formalized expertise in acts".

The most frequently observed schemes relate, for example, to apprenticeship programs, in which professional learning is carried out by alternating work periods and training periods. The engineering of apprenticeship programs aims to analyze work to design training situations related to the practice of work, taking into account the challenges of temporality and synchronization between the experience lived at work and the training experience. Other measures frequently observed relate to the observation-analysis of practices but also, and more recently, to training actions at the workplace (particularly in France since 2018). Here, the very practice of the work being done can be a source of professional learning, as long as it is carried out in a context where new elements are introduced and regular "debriefing" moments and an analysis of said activity are organized, provided that learning is associated with a third party on-the-job (see Chapter 7). The diagram is as follows: work sequence, debriefing/analysis moment with a third party, new work sequence, new debriefing/analysis moment, etc. Finally, we can also mention support situations (tutoring, for example, enabling the trainee to take a step aside) and professional action-research situations placing a professional in a research posture (alone or with others) regarding the activity in progress, as a way of generating relative professional learning to experiment with other possibilities of the activity.

Table C.2 summarizes professionalization paths and the associated work and/or training schemes.

	Learning developed (dominantly)	Expertise produced (dominantly)	Frequently associated schemes
Professionalization through formal training	Formal knowledge in the sense of "putting into words"	Formal expertise often removed from concrete and unique work situations	Off-the-job training
Professionalization through work	Knowledge of action or experience in the sense of "putting into action"	Expertise in actions often closely attached to work situations	Work practice Training actions at the workplace Simulation or role play Project groups or problem-solving groups
Professionalization through the articulation of work-training	Knowledge about and for action in the sense of "putting actions into words"	Formalized expertise in acts	Apprenticeship programs Observation-analysis of practices Training actions at the workplace Accompaniment Professional action-research

Table C.2. *Professionalization paths and the associated work and/or training schemes*

Overall, there appear to be many different logics associated with each professionalization direction or, put differently, with each of the three forms of professionalization. This underpinned the choice to organize the encyclopedia in function of the three meanings of professionalization, making it easier to systematically present the available knowledge produced in disciplinary fields which are distinct and often have little exchange between one another. This structuring and presentation can lead the reader to think that professionalizations operate separately, but, as we will show in the following section, this is not the case, leading us to plead for a communication of the knowledge produced *by* and *within* each of the disciplinary fields.

C.2. Beyond the apparent "boundaries", the observation of an overall movement reflecting an interdependence of professionalizations

Far from constituting separate logics each functioning independently, these three meanings actually refer to interrelated dynamics characterized by "debates", "tensions" and "reciprocal transformations" in environments, organizations, collectives and individuals. In many contemporary situations, "professionalization-professionality" receives the impact of "professionalization-professionalism", which in turn has an effect on "professionalization-profession" logics. The essential challenge, the determining question comes down to the control of these heterogeneous processes: who is in control?

C.2.1. *A priori, different processes and conceptualizations argue in favor of the existence of separate and autonomous logics*

As for the dynamics observed and the main actors concerned, the previous chapters seem to show that the actors do not experience the different challenges present in the three forms of professionalization in the same way: while organizations express the rise in professionalism forms, individuals are in search of professionality. Finally, the collectives seek affirmation and recognition of a profession in the case of professional groups, and strive for the recognition of their own professionalism for organized work collectives when workers have had to actively engage in profound reorganizations of work.

Now, as for the conceptualizations and lines of research present, each of these logics seems to be studied by a different disciplinary field producing its own conceptualizations, without being truly related to the others and implying that they translate separate logics. While the sociology of professions and the sociology of professional groups shed light on "professionalization-profession", the sciences of work and organizations study "professionalization-professionalism" and training sciences are more interested in "professionalization-professionality".

C.2.2. A close articulation between the three professionalizations based on bonds of an agonistic nature

Often, talking about professionalization actually refers to a "debate" between at least two projects: a project for change, expressed by socioeconomic environments and organizations (professionalization project for others), and a project for professional development and recognition promoted by professional groups, organized work collectives and individuals. In other words, professionalization would involve a relationship between:

– the intentions, challenges and rhetoric of the political, socioeconomic and organizational environments (companies, associations, public ministries) due to the developments which directly concern them or which they provoke;

– challenges, rhetoric and positioning strategies from social groups and individuals based on a logic recognizing specificities.

In the contemporary period, there is an overall coherence between developments in public environments (New Public Management) and private (new, more flexible forms of work organization), changes in human resources management methods, as well as changes in training forms. Everything seems driven by the valorization of an individual who is more "autonomous and responsible in his life and at work", as well as the search for greater overall efficiency at work and in social life which often clashes with the challenges of professional affirmation and recognition expressed by individuals and groups, which Osty (2003) described as the "desire for a métier".

Let us develop this point:

– On the one hand, with regard to sociopolitical environments and organizations, we see a movement revealing converging challenges, bringing forward a concept of "professionalization-rise in professionalism forms". Since the beginning of the 1980s, the promotion of political (decentralization laws), social (discourse on the active citizen) and organizational decentralization (expectation of an employee capable of initiative and decision) has accompanied an intention for greater empowerment and responsibilization of individuals (call for more "transversal" skills, reference to "accountability"). This signals an implicit intention for an "ongoing movement" of people ultimately adapting to a new mode of "governance" (the main orientations being fixed centrally, its implementation being delegated but more controlled on the basis of criteria defined at the level of decision-makers in major orientations), targeting greater overall efficiency for social life and work. But the situation is far from being free of contradictions or paradoxical injunctions. The autonomy-responsibility model is often strictly framed by the rise in evaluative standards at work which are renewed and clarified very quickly: we have to be autonomous and gifted, display initiative and creativity, but respect a more stringent work framework. In any case, these analyzes lead to revisiting the classically advocated conception of a professional: they can no longer only be, as in the past, an expert endowed with expertise whose contours are stable (and therefore easy to state), but must also be an expert capable of moving expertise (and that of the métier on a collective level) because they are invited not only to "know how" but also to "know how to analyze what they do", to "know how to adapt what they do" depending on situations and to "know how to communicate what they do and the adaptations" implemented, to be the person of "the" situation(s) ("know how to do *that*").

– On the other hand, with regard to individuals and collectives, we see a strong demand for recognition relating to what we have called "professionalization-profession" and "professionalization-professionalities". The challenge here is to obtain a better place and greater autonomy in the space of professional activities. Let us take the example, in France, of the regular demands for autonomy of nursing professionals in relation to doctors, a situation which has already been settled in most English-speaking countries. The question is to constitute a counter-power in organizations against their tendency to take control of the definition of tasks and the profession's contours. Thus, when versatility is strongly prescribed and causes an identity crisis (Dubar 1991), the quest for professionalization aims to find meaning at work and stability, as a guarantee of recognition. Finally, it also involves the recognition of professionality developed throughout work practice (as the workers' demands in the Fiat company in Italy at the end of the 1960s: *professionnalita* vs. *qualifica*).

As the following diagram attempts to indicate, this analysis leads us to believe that there are strong links between the projects of organizations, the projects of collectives and those of individuals. On some occasions, these links reflect convergent logics. On the contrary, more often than not, they reveal strong divergences (especially in contexts of change), leading to thinking of "professionalizations" from the angle of regulation processes and social transactions between actors expressing different intentions.

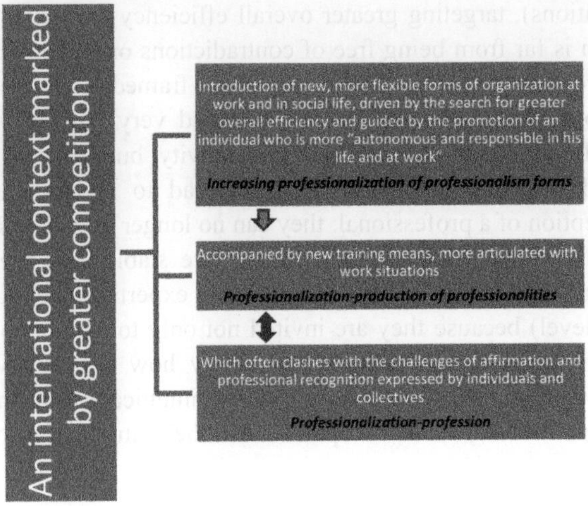

Figure C.1. *Links between the projects of organizations, the projects of collectives and those of individuals. For a color version of this figure, see www.iste.co.uk/demaziere/encyclopedia.zip*

C.2.3. *Leading to "reciprocal and mutual transformations" of environments, organizations, collectives, individuals in multiple space-times*

As we will see, these (agonistic) relationships between different projects are accompanied – through and in their existence and deployment – by reciprocal transformations of the actors involved (organizations, collectives and individuals).

Let us take two examples:

– The Canadian textile company mentioned previously was under the influence of strong competition in its environment which had the effect of implementing a rapid and large-scale change project, aimed at transforming the way workers worked. The scheme for analyzing work and defining new ways of working conducted by the workers themselves unexpectedly created new solidarity and a strong affirmation of their new role in the company (they wanted to be listened to, since they had become specialists in their work). In turn, in a manner not anticipated by the company's management, this had destabilizing effects on management. The company then made new decisions aimed at repositioning the role of managers. This example reveals the existence of close links between the environment (strong pressure from competition), the organization (the implementation of a rapid change project to face such competition) and the workers (the development of new solidarities, the affirmation of a collective and a new expertise), which are accompanied by "reciprocal and mutual transformations": the workers unexpectedly become a collective acting in the company, the company is driven to transform, beyond its initial intentions, the prerogatives of its management and the environment sees the re-emergence of a competitor on the textile market.

– The above-mentioned sector of direct employment at home constitutes another example in which the starting point of the dynamic of professionalization observed is not the environment, but the conviction and will of a handful of actors (individual employers) trying to initiate a strategy for the social organization of the sector through institutional negotiation and recognition of means of remuneration for salaried activity at home, through the definition of home-based professions, the construction of professional certifications and the proposal of training which little by little becomes a means for the identification and work recognition of thousands of relatively "invisibilized" employees. As from that moment, home employees began to organize themselves collectively and to ask in return for an improvement in their remuneration, which led to thinking about a more precise classification grid, also requiring the implementation of an ongoing training offer. This example also shows that professionalization sets up a system of links, which is accompanied by "reciprocal and mutual transformations".

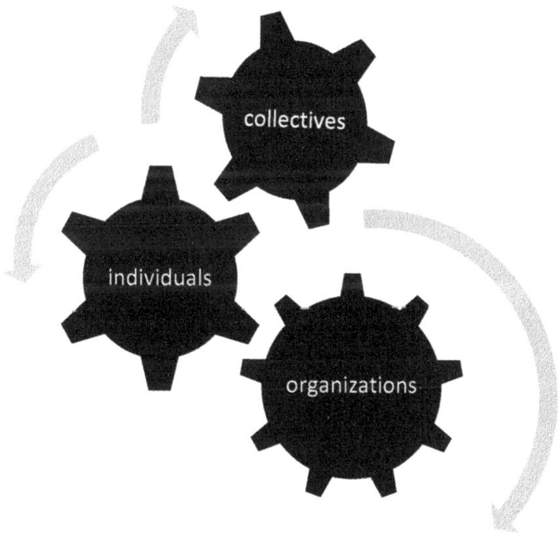

Figure C.2. *Close links and reciprocal transformations between environments, organizations, collectives and individuals*

In a nutshell, these two examples lead us to consider that "professionalization" refers to:

– On the one hand, to configurations (Elias 1987) in the sense of systems of close links and reciprocal transformations between environments, organizations, collectives and individuals, as will be shown in the following diagram. These are cross/joint/mutual/reciprocal transformations driven and piloted by inseparable challenges involving relentless change in organizations, collectives, individuals and places, therefore in identity (places prescribed, claimed, assigned at the level of organizations, groups, individuals).

– On the other hand, transformations which take place in multiple temporalities; the short time of the work or training scheme, the long time of the individual's trajectory as well as:

> The biographical time of individual career paths, subjective time of anticipations of the future, discontinuous time of the maturation of professional projects, limited time of training situations, regulated time of the accomplishment of tasks, uncertain time of managerial strategies, the rationalized time of work organization, the instantaneous time for exchanges and interactions, the sedimented time for the crystallization of work collectives, etc. (Demazière et al. 2012, p. 10).

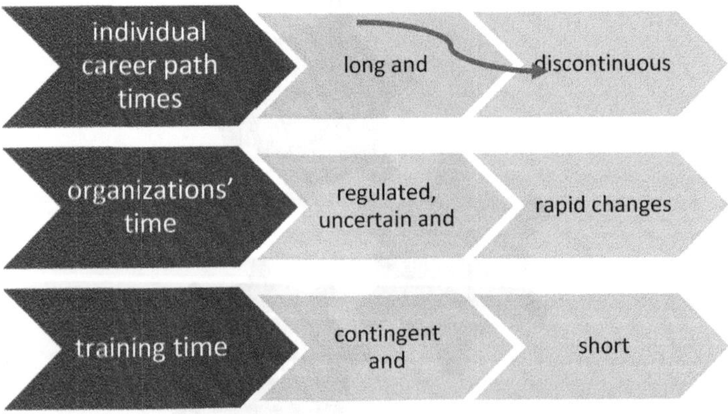

Figure C.3. *Transformations which take place in multiple temporalities. For a color version of this figure, see www.iste.co.uk/demaziere/encyclopedia.zip*

One of the strong contemporary characteristics seems to be the acceleration of prescribed and organized socio-professional times, as shown by the work of Rosa (2010), which indicates that the major experience of modernity is that of acceleration: "everything is always getting faster", accompanied by stress and a lack of time. The references to the fragmentation of time (Urry 2000), its acceleration (Rosa 2010) and the generalization of the temporality of emergency (Laïdi 1999) are the subject of a fairly broad consensus. It allows us to think that professionalizations are part of a context of greater urgency, as mentioned by Datchary and Gaglio (2014, p. 5) "because time, as if caught in a vice, is compressed and compresses, a fast-paced temporality supplants the others: the urgency. Thus, the first phenomenon that attracts the attention of researchers lies in highlighting the 'emergency pathologies' the modern worker suffers from". Therefore, these relationships of reciprocal and mutual transformations of individual, collective and organizational actors involved in professionalization can be accompanied by conflicts of temporalities, tensions and contradictions, generated by the overlap of plural and nonhomogeneous temporalities. Alter (2000, 2003) proposes the concept of "dyschrony" to express these conflicts, observing that, in the business environment, many change projects overlap at the same time (e.g. introducing just-in-time and quality certifications). One of the characteristics of permanent change in many sectors is that it generates conflicts of temporality: "one has to make changes continuously, while protecting the skills and routines that enable efficiency" (Alter 2003, p. 7). Changes are of different natures and in an ongoing manner, the organization becomes a continuous movement.

C.3. In conclusion: some blind spots and future challenges regarding the study of professionalization

The analyses presented in this encyclopedia constitute an organized state of the work done on professionalization from different disciplinary points of view. They allow us to better grasp the challenges and forms that these professionalizations take by describing a field initially marked by a certain complexity due to the multiplicity of actors and issues present.

These available analyses would undoubtedly deserve to be supplemented with more in-depth investigations on the following dimensions:

– In relation to the previous point: if the acceleration of socioprofessional and institutional times is a strong trend of the contemporary period, would not it be useful to deepen the analysis of the different temporalities present in the dynamics of professionalization, so as to better understand not only the historicity of the phenomena, but also their rhythm? More particularly, one of the challenges would probably involve better seizing the articulations or conflicts (like the above-mentioned dyschronies) between multiple temporalities (that of individuals, institutions, social groups, etc.). While the description and analysis of each of the temporalities is often available, independently of one another, the unique process of articulation of the different temporalities that intersect has been barely studied. An approach based on the varied temporalities and rhythms of professionalization would still need to be developed, even if research has already started to introduce it (e.g. Alhadeff-Jones 2014; Roquet 2018).

– If we consider that professionalization is primarily a prescription, affirmation or even social recognition (of skills, expertise, status, etc.), can we better understand the evolution of the modalities of prescription, affirmation and of recognition in the dynamics of professionalization, by introducing an often "blind spot" actor, who is the client or the user (and relatives: caregivers, peer caregivers), in particular for activities and professions addressed to others, for which the notion of co-activity is central? For example, to what extent does the fact of giving a different place in the social policies of many countries to users from social work professionals ("placing the user at the center") does not lead to transforming the usual spaces of prescription, but also of recognition of the work of social professionals (the user themself becoming an agent of prescription and recognition of the professional's work and invited to report on it)? In other words, we are perhaps seeing new and diversified forms of prescription and recognition of professionalization beyond the initiatives usually taken by institutions. At the same time, this invitation to a user-prescriber is accompanied by a search for a certain form of user professionalization (and of their loved ones with paid and trained caregivers). The person is called for example to act as a "professional", as in the therapeutic education of the patient, the social education of the user of social work, the imposition of preventive self-therapy on citizens with regard to Covid, for instance.

– Finally, one of the challenges of future analyses with a view to better understanding professionalization is to pursue research articulating the perspectives initiated in this book, particularly considering the fact that the realities studied under the term "professionalization" relate of multiple combinations which transform over time. In this view, professionalization is above all a matter of composite processes. A new theoretical but also methodological ambition appears: to develop not only places and moments favoring the crossing of available theoretical grids relating to different disciplinary approaches (sociology, education and training sciences, social and work psychology, management sciences, etc.) on the same object, but also thinking about mixed methodologies, articulating synchrony and dyschrony at different scales, singular and regular analysis of the activity of individuals, organizations and customers and users, "transactions"/debate/organizational-subject-collective struggles, etc.

To better seize these dynamics, it requires crossing insufficiently linked work traditions, for example:

– For the study of transformations of environments, organizations and the analysis of the dynamics of "collectives", we would have to better articulate certain works in the sociology of organizations (e.g. the strategic approach and game theory) with a critical sociology of work and developments in training and critical approaches to broader social and political developments.

– For the study of the transformations of individuals, it would be useful to closely articulate research in work analysis and work ergonomics and clinical work with interdisciplinary approaches regarding the links between activity (work or not) and the construction of situated learning, without forgetting interactionist approaches to identity construction/transformation.

The challenge is to seize the relationships between individuals, collectives, organizations and environment, which requires placing future work in a movement that goes beyond multidisciplinarity to open up to interdisciplinarity in the sense of collaboration between disciplines capable of taking into account at the same time these different facets currently studied in relatively separate ways. One of the major challenges related to this search for interdisciplinarity lies in the attempt to achieve an overall understanding capable of thinking in the same movement (and not separately), the co-transformations produced by social transformations, and the transformations of organizations, social groups and individuals, without forgetting the transformations of clients and users for activities addressed to others.

C.4. References

Alhadeff Jones, M. (2014). Pour une approche réflexive et critique des rapports entre temporalités et professionnalisation. *Phronésis*, 3/4, 4–12.

Alter, N. (2000). *L'innovation ordinaire*. Presses Universitaires de France, Paris.

Alter, N. (2003). Mouvement et dyschronies dans les organisations. *L'année sociologique*, 2, 489–514.

Datchary, C. and Gaglio, G. (2014). Hétérogénéité temporelle et activité de travail. Entre conflits et articulations. *Revue d'anthropologie des connaissances*, 8(1), 1–22.

Demazière, D., Roquet, P., Wittorski, R. (2012). *La professionnalisation mise en objet*. L'Harmattan, Paris.

Dubar, C. (1991). *La socialisation, construction des identités sociales et professionnelles*. Colin, Paris.

Elias, N. (1987). *La société des individus*. Fayard, Paris.

Laïdi, Z. (1999). *La tyrannie de l'urgence.* Fides, Montreal.

Osty, F. (2003). *Le désir de métier : engagement, identité et reconnaissance au travail.* PUR, Rennes.

Roquet, P. (2018). Temps long/temps court en situation formative et professionnelle : les effets sur les constructions identitaires. *Education Permanente*, 217(4), 33–41.

Rosa, H. (2010). *Accélération, pour une critique sociale du temps*. La Découverte, Paris.

Urry, J. (2000). *Sociology Beyond Societies: Mobilities for The Twenty First Century.* Routledge, London.

Wittorski, R. (2007). *Professionnalisation et développement professionnel.* L'Harmattan, Paris.

Wittorski, R. (2016). *La professionnalisation en formation*. PURH, Rouen.

List of Authors

Stephen BILLETT
Griffith University
Brisbane
Australia

Valérie BOUSSARD
Printemps
Paris Nanterre University
France

Florent CHAMPY
LISST
CNRS
University of Toulouse
France

Didier DEMAZIÈRE
CSO
CNRS
Sciences Po
Paris
France

Sandra ENLART
Apprenance Team
Paris Nanterre University
France

Mokhtar KADDOURI
CIREL Laboratory
University of Lille
Villeneuve d'Ascq
France

Philippe MAUBANT
CRIFPE
University of Sherbrooke
Canada

Arnaud MIAS
IRISSO
Paris-Dauphine University
France

Joris THIEVENAZ
LIRTES
Université Paris-Est Créteil
France

Linden WEST
Canterbury Christ Church University
UK

Richard WITTORSKI
CIRNEF Laboratory
University of Rouen
France

Index

A, B

C, D

E, F

H, I, J

K, L

M

N, O